ENERGY AND ELECTRICITY IN INDUSTRIAL NATIONS

Energy is at the top of the list of environmental problems facing industrial society, and is arguably the one that has been handled least successfully, in part because politicians and the public do not understand the physical technologies, while the engineers and industrialists do not understand the societal forces in which they operate. In this book, Allan Mazur, an engineer and a sociologist, explains energy technologies for nontechnical readers and analyzes the sociology of energy.

The book gives an overview of energy policy in industrialized countries including analysis of climate change, the development of electricity, forms of renewable energy, and public perception of the issues. Energy is a key component to environment policy and to the workings of industrial society. This novel approach to energy technology and policy makes the book a invaluable interdisciplinary resource for students across a range of subjects, from environmental and engineering policy, to energy technology, public administration, and environmental sociology and economics.

Allan Mazur, an engineer and a sociologist, is Professor of Public Affairs in the Center for Environmental Policy and Administration at Syracuse University, USA.

ENERGY AND ELECTRICITY IN INDUSTRIAL NATIONS

The Sociology and Technology of Energy

Allan Mazur

LONDON AND NEW YORK

First edition published 2013
by Routledge
2 Park Square, Milton Park, Abingdon, Oxon, OX14 4RN

Simultaneously published in the USA and Canada
by Routledge
711 Third Avenue, New York, NY 10017

Routledge is an imprint of the Taylor & Francis Group, an informa business

British Library Cataloguing in Publication Data
A catalogue record for this book is available from the British Library

Library of Congress Cataloging-in-Publication Data
Mazur, Allan.
Energy and electricity in industrial nations: the sociology and technology of energy/Allan Mazur. – First edition.
pages cm
Includes bibliographical references and index.
1. Energy consumption – Social aspects – United States. 2. Power resources – Social aspects – United States. I. Title.
HD9502.U52M295 2013
333.790973 – dc23
2012038235

ISBN13: 978-0-415-63441-0 (hbk)
ISBN13: 978-0-415-63442-7 (pbk)
ISBN13: 978-0-203-09448-8 (ebk)

Typeset in Times by
Florence Production Ltd, Stoodleigh, Devon

Printed and bound in Great Britain by MPG Printgroup

For Max, Wren and Marlo

CONTENTS

FIGURES

TABLES

PREFACE

There were no industrial societies 200 years ago, and there was barely any use of energy by today's standards. Electricity, in the form of current flowing through a wire, did not exist. Living life was a constant struggle with the environment, but there were no "environmental problems" of the kind that occupy today's environmental organizations. The topics of this book begin in the nineteenth century, when a small group of nations industrialized, began burning fossil fuel for their energy, and, as the twentieth century approached, started generating electricity to produce light and heat and to run electric motors.

Engineers and sociologists

Engineers, inventors, entrepreneurs, business people – all "practical men" – were involved from the start of industrialization, with a common goal of making money. There was little contribution from academic scientists, those who pursue pure knowledge rather than profit. Michael Faraday was an exception, discovering in his laboratory that electricity could be produced by moving a wire near a magnet, though at the time there was no practical use for this phenomenon. Until steam engines and electricity were well established, science contributed less to the practical arts of energy production than practice contributed to pure science, as when Sadie Carnot famously laid the basis of thermodynamics by analyzing the efficiency of steam engines.

Least involved were the intellectual forebears of those today called "sociologists," if for no other reason than the difficulty for humanities-oriented scholars to understand how the practical men made their devices work, especially the seemingly magical force of electricity. (I speak of "sociologists" broadly, not excluding other social scientists, and these are no longer as oriented to the humanities.) Even today, while the topics of energy and electricity are well treated in economics, they remain outside the usual scope of other social sciences. There is no "sociology of energy," which seems odd in view of the central role that energy and electrification play in modern society and politics.

We need an interdisciplinary approach to the problems of technological society, involving not only those trained in engineering and science, but others coming from the humanities, law, and social sciences, with some seasoning from "ordinary" citizens, environmentalists, journalists, and even politicians. There has not yet been much productive movement in this direction. I'll wager that out of 100 nontechnical professionals or politicians, not five know how a piece of coal is converted to electric current, or why that current can toast a slice of bread, much less run a computer. With broad-based knowledge so scarce, how is a modern nation to reasonably deal with foreign oil dependency, climate change, or the possibility of disasters caused by nuclear reactors or deep-sea oil drilling?

This small book is no cure-all, but it should build a bridge between sociologists who know little of technology, and the engineers and natural scientists who at times seem incredibly naïve about the behavior of citizens and the workings of society.

Industrial nations and the Third World

My focus on industrial nations ignores most of the world's population, those in nonindustrial nations and often called the "Third World," a remnant from the Cold War when the US and its industrialized allies were the First World, the Soviets and their allies the Second World, and everyone else the Third. The "first" and "second" designations had fallen from use even before the collapse of the Soviet Union. The Third World persists as a less derogatory label than others that have been used for nonindustrial peoples, the majority of humanity, and unfortunately ignores their diversity. Parts of the Third World are essential for our analysis of the industrialized world. Certainly for global energy consumption, pollution, and climate change, one must include China in the discussion, and also India. For petroleum supply, resource wars, petrodictatorships, and terrorism, one must include the OPEC nations and other major oil producers of the Third World. Still, this book is about the industrialized world, with an admitted slant toward the United States, which is the most important nation in this story, certainly in the consumption of energy and electricity, using for example one-quarter of the world's annual oil production. For good statistical overviews of energy in the world as a whole see the most recent editions of the *BP Statistical Review of World Energy* (http://www.bp.com/statisticalreview), the International Energy Agency's *World Energy Outlook*, and for the United States specifically, the US Energy Information Administration's *Annual Energy Outlook* (http://www.eia.gov/forecasts/aeo/).

The industrial nations, or alternatively "developed" nations, are mostly located in Europe, North America, and Japan. It is difficult to ignore the inference that industrial nations are more modern, advanced, progressive, somehow better than agrarian societies. This is a touchy subject that I intend to avoid, except to acknowledge that there surely are virtues of traditional society that have been lost during industrialization. On the other hand, looking around the globe at Third World populations, many of them are striving to industrialize – often in their own way rather than emulating Western values, while I see no industrial nation seeking a reversion to preindustrial agrarianism. So an image of societal evolution, from agrarian to industrial (to postindustrial?), is not completely misplaced.

What exactly counts as an industrial nation? Generally, it is a country that within the past two centuries has adopted epochal changes in technology, energy supply, wealth, education, and social organization, setting it apart from agrarian societies of the past and present. A common feature is the small proportion of people directly involved in growing food, which traditionally was nearly everyone, and in today's Third World still occupies half or more of the population. Some industrial nations minimize their agricultural sector, importing what they need, while others still grow most of their food but do it far more efficiently than in the past, replacing human labor with technology and energy from fossil fuels. History's most prolific producer of food, the United States, has only about three percent of its population directly occupied in farming, most of the rest living in cities or suburbs. Other common features of industrial nations are high income (of course, there are still poor people) and high use of energy and electricity. In recent years the industrial nations have tended to have gross national income *per capita* of above $25,000 (expressed in US dollars and purchasing power parity), and usually consume over 120 gigajoules of primary energy per capita, of which 15 percent or more is supplied to end users as electricity. These are dry numbers without much meaning to most readers, but it will suffice to say that all are high numbers by Third World standards.

Energy and power

We must discuss definitions and enter the thicket of units of measurement. *Energy* is the ability to do work, and it comes in many forms: as heat (thermal), light (radiant), motion (kinetic), electricity, nuclear energy, and gravitational energy. Energy may be stored, for example in the chemical bonds of a lump of coal, or in the potential energy of a weight suspended high above the ground. It can be liberated by burning the coal, or by dropping the weight, thus converting one form of energy into another, or by applying energy to do physical work.

Time is not involved in the definition of energy or work. But sometimes it is important to know how fast the energy conversion occurs, or the work is done. *Power* is the rate at which energy is converted from one form to another, or is applied to do work. (The relationship between power and energy is analogous to that between speed and distance covered.) The *watt* is a familiar unit of power. If a 60-watt incandescent light bulb stayed lit for one hour, it would consume (or convert) 60 *watt-hours* (Wh) of electrical energy, radiating some of that as visible light and much as heat. (The Law of Conservation of Energy guarantees that the amount of electrical energy going into the bulb is equal to the amount of energy leaving the bulb as heat and light.) The heat portion would be wasted energy unless one's intent was to heat a room as well as light it. By the way, many museums have exhibits that allow the visitor to peddle a stationary bicycle connected to a generator in order to power a conventional light bulb. Most of us have to peddle at maximum speed to get light from the bulb.

Historically, researchers and engineers studying each form of energy or power have named their own units. The unit *horsepower* was originally intended to compare the power output of the newly invented steam engine with that of a draft horse, as in "this engine will draw as much water from the mine as a sturdy horse can do." The comparison alluded to a short time period, a rate of work, not to the total amount of work done in extracting water during a full workday. (A one horsepower engine, or horse, operating for an eight-hour workday would use eight horsepower-hours of energy in extracting water from the mine.) Despite varying definitions of the horsepower, this archaic unit is still favored when rating engines. Today one horsepower is defined precisely as 745.699872 watts. Again, horsepower and watt refer to power, the rate at which energy is used, not to a total amount of energy that is used.

Definitions of units were often inexact and variable from place to place. Obviously the British invented the venerable *British thermal unit* (Btu), approximately the amount of energy needed to heat one pound of water from 39° F to 40° F. The Btu is still used in some industries, and its multiple, the Quad (= 10^{15} Btu), is still used in some national energy accounts, resisting replacement by metric units.

Each fuel has its own units: tons of coal, barrels of oil, cubic feet of natural gas. The "short ton" of coal used in the US to mean 2,000 pounds is lighter than the "long ton" traditionally used in Britain (2,240 pounds). Furthermore, the ton is actually a unit of weight, not of energy, because the chemical energy in a quantity of coal is highly variable, depending on its quality.

This topic would be hard even if we restricted ourselves to the rational metric system, but with the United States stubbornly refusing to adopt metric measurement (except for money), there are both metric and nonmetric units for nearly everything. Just to keep your head spinning, the metric unit *tonne* differs from both the US and UK nonmetric tons. Today international accounts often use *tonnes of oil equivalent* (toe) as a unit to compare or combine the energy in different fuels. Since a tonne of oil from one place would not generally have the same energy content as another tonne of oil, the toe is somewhat arbitrarily defined as 41.868 gigajoules (gJ), a metric unit of which I am especially fond. All energy units, however familiar or strange, can be more or less converted into joules, thus one watt-hour = 3600 J; one ton of TNT (equivalent energy) = 4.184 gJ.

James Prescott Joule was an English physicist and brewer who discovered the relationship of heat to mechanical work, leading to the Law of Conservation of Energy. His name is given to the basic unit of energy in the International System of Units (ISU), a form of the metric system devised in 1960 to rationally simplify the morass of physical units. Watt is the unit of power, named after James Watt, who appears in Chapter 1. As usual in the metric system, notation uses powers of ten, e.g., one kilowatt (kW) = 10^3 W; one megawatt = 10^6 W; one gigajoule = 10^9 J; one petajoule = 10^{15} J). I will mostly use the ISU system, but I admit that other units have crept into this book because the force of habit is strong, and in some instances traditional units will be more meaningful to most readers.

One might think that adoption of ISU would make the business of energy accounting simple, but complications remain. Two authoritative sources for international statistics are the US Energy Information Agency (EIA) and the International Energy Agency (IEA), the latter an autonomous organization of 28 member countries, mostly the industrial nations. For the most part they agree on how to tally the amount of primary energy consumed by a nation, but they differ on how to count the amount of primary energy devoted to producing electricity. (Electricity is a secondary form of energy because usually it is generated by burning primary fuels like coal or natural gas.) For a typical fossil-fueled electrical generator, about three units of primary energy must be burned to produce one unit of energy in the form of electricity. But no fuel is burned when producing hydroelectricity from water flowing through the generators in a dam, so how should one count the "primary energy" that produces hydroelectricity? The IEA assumes for accounting purposes that the amount of "primary energy" going into the dam is equal to the amount of electricity coming out of it. The US EIA, in contrast, using the analogy of a fuel-burning plant assumes that the amount of "primary energy" going into the dam is three times the output (Lightfoot 2006). For a nation like Norway, which uses a lot of hydroelectricity, primary energy consumption is far higher by EIA accounting method than by IEA accounting. The important take-away point is that there may be large uncertainties in energy statistics, so one should not attribute much importance to small differences between numbers.

Structure of the book

The industrial energy "system" is, both technically and sociologically, a hodgepodge of disparate and often conflicting actors, fuels, technologies, government regulations, business practices, public controversies, and sometimes disasters. To make this book coherent, I have grouped topics into five parts.

Part I, "The big picture," describes the industrial transformation of the past two centuries, which is the ultimate cause of today's energy problems. As antidote to the popular view that *the* energy problem is fuel depletion, or perhaps global warming, I set these among the other, possibly worse, troubles that we suffer by using so much energy. I give a sociologically relevant overview of the flow of energy and electricity through an industrial society, emphasizing how this differs from preindustrial societies. Closing this section, I examine the degree to which the highest-consuming industrial nations benefit from their use of energy and electricity.

In Part II, "Energy sources and consumption: using more, and more, and more . . . ," I describe the major sources from which we obtain energy, first the fossil fuels (coal, oil, natural gas), then the non-carbon sources (nuclear, renewables from the sun, geothermal). I address the central Malthusian question: How much does population growth account for our increasing consumption of energy and electricity?

While electricity as a distinctive form of energy is discussed throughout the book, it is the central focus of Part III, where I explain the operation of national power grids, how they grew in America after Edison's invention of the light bulb, and the consequences for Japan's electrical system of the tragic earthquake and tsunami of 2011.

Part IV, "Energy controversies," is the most traditionally sociological section of the book. I first tell how "rational" engineering approaches to projects may conflict with, and make less sense than, the seemingly ideological or emotional objections of project opponents. I describe the common social structure of technical controversies, whether their focus is global warming, "fracking," or the siting of energy facilities. And I describe the important role of the mass media in the escalation or de-escalation of controversy.

Finally, in Part V, "Progress and regress," I explain and evaluate some of the energy solutions that have been proposed or implemented since the oil shocks of the 1970s shook the industrial nations out of their complacency, and I propose guidelines for future energy policy, to which I hope readers of this book will contribute.

Acknowledgements

I appreciate the help or advice of Jessica Bian, Steven Brechin, Jack Casazza, John Casteel, Lee Clarke, Matt Coulter, Paul Hines, Emma Krupnick, George Loehr, Rachel Mazur, Todd Metcalfe, Jerry Miner, David Nevius, David Owen, Michael Shellenberger, Donald Siegel, Donald Shaw, Julie Mazur Tribe, Austin Troy, Peter Wilcoxen, and Robert Wilson. Of course, none of them necessarily endorses positions stated here.

PART I

The big picture

1

THE AGRARIAN AND INDUSTRIAL TRANSFORMATIONS

People more or less human, members of the genus *Homo*, have been around for only about two million of the earth's 4.5 *billion* years. Modern-looking humans, *Homo sapiens sapiens*, appeared about 200,000 years ago. During nearly all the time since, nothing very interesting happened from the standpoint of a sociologist or an engineer. People continued living in small groups, occupying temporary living sites, hunting and fishing or gathering naturally growing food. Their cultures were static, people in one millennium doing the same things, living nearly the same kind of lives, as people did in the last millennium or would in the next, using simple technologies that changed slowly.

Ocean barriers that today look impassable with primitive conveyances were more easily crossed when sea levels were lower because glaciers of the last ice age held a large portion of the earth's water. By 10,000 years ago, modern-looking humans had migrated by foot or boat from their original habitat in Africa to occupy all of the continents inhabited today. Still they all lived in small societies of collectors and hunters.

The agrarian transformation

Beginning 10,000 years ago, there was a profound change in the human condition. Life in several places was transformed into an agrarian mode, with people settling in permanent communities supported by nearby fields of grain and by animal husbandry. The animals became sources of power and transport as well as food and pets. The populations of growing towns became differentiated into separate classes, one better off than another, with some form of king holding control, partly through hereditary right and partly through the strength of military alliances. Cultures grew, merged, diffused, and diversified.

The number of people grew much larger than ever before, though still very small compared to today's population. Agrarian food sources fed more people, and agrarian people produced more food than was available on the natural landscape. Like the chicken-and-egg problem, it is fruitless to ask if one caused the other. We do not know why the agrarian transformation occurred. Because of a coincidence in timing, it is commonly surmised that agrarian innovations were somehow triggered by the warming climate that ended the last ice age about 10,000 years ago, opening new ecological opportunities for social change.

This new agrarian life was the base upon which civilization would emerge within another five or six thousand years. I use the word "civilization" as archeologists do, meaning an advanced form of agrarian society that usually has pottery, writing, calendars, astronomical observation, mathematics, monumental architecture, urban communities with thousands of inhabitants, planned ceremonial and religious centers, specialization in arts and crafts, metallurgy, and intensive irrigation projects – not necessarily all of these, but most.

At first archeologists thought that agrarian civilization arose uniquely in Mesopotamia and then spread via human migration or cultural diffusion to other geographical centers, as if a random spark had ignited a conflagration. But research during the past half century convincingly shows that these cultural innovations occurred independently in at least six places: Mesopotamia, Egypt, India, China, Mexico, and Peru. In each pristine area, the old hunter-collector life changed into an advanced agrarian civilization indigenously, without important influence from other major centers. That this should occur in at least six places during a "brief" period (five or six millennia), after 200,000 years of relative stasis, seems miraculous. A fanciful writer of the 1960s named Erich von Daniken (1968) argued in a bestselling book that the coincidence could only be explained by the visitation to earth of advanced visitors from another world who sowed the seeds of civilization. It was an absurd claim that does not stand up to critical examination, but one that highlights the remarkable scope and suddenness of this transition.

If extraterrestrials had occasionally visited earth during the agrarian period, they would have found one human civilization pretty much like another. Nearly every person, young and old, was engaged in raising crops and livestock. Cities and towns, the centers for commerce and manufacture, were interconnected by trade routes, some quite long, where land travel was by foot or animal, and sea travel was by sail or oar. Communication moved no faster than a person could. Manufacturing was carried out by individuals or small groups of people using handcraft methods, with flowing water, wind, or fire as their source of inanimate power. Large monuments and ceremonial buildings were constructed of stone, wood, and dried brick, while metal was reserved for smaller objects. Land was the most valued resource, and even the richest people rarely had large (by modern standards) amounts of spendable capital. A hereditary monarch usually stood at the top of a rigid status hierarchy, supported by a favored aristocratic class (often warriors), and at the bottom was the poor mass of peasants. Often, one's position in this hierarchy was fixed at birth and tied to a specific piece of land, as in feudal Europe and Japan, but sometimes careerists were assigned to distant posts, as in the administrative bureaucracies of Imperial Rome or China. In either case, the rights and obligations of one rank toward another were well specified (always to the advantage of the higher ranked), and an unquestioned religion and its clergy justified the existing arrangement as right and proper. Armies depended upon blades and animal power. Each civilization had impressive (by modern standards) artistic achievements.

Europe of Columbus's time was not advanced economically, technologically, or aesthetically over some other parts of the agrarian world. The Forbidden City of the Chinese emperors, the Islamic world of the Moors, and capital cities of the Incas and Aztecs were as grand as anything in Christendom. Some agrarian societies were less impressive, but in their fundamental operation, certainly in their use of energy, none was very different from others of prior millennia.

If the extraterrestrials returned in 1900 for another reconnaissance, they would have been surprised at the changes. By then European civilization was preeminent, followed by its direct descendant, the United States. No longer agrarian, these societies – and to a lesser extent Japan – had become industrial civilizations, their populations moving from farms to cities, taking jobs in manufacturing and commerce. Efficient factories employing large workforces, using steam power and even electricity, produced goods in far greater quantity, and often cheaper, than ever before. Finished products were sold around the world, transported by trains and steamships that returned with raw materials to feed the factories. Steel was a common building material, and electrical communication was instantaneous. Money and industrial resources had replaced land as the most valued form of capital, with many individuals and corporations having accumulated immense amounts. Some of Europe's powerful monarchies had fallen (others would soon follow) along with special privilege for their aristocracies; America was a democracy where all white males had the same rights and obligations. Catholicism was in decline in Europe, challenged not only by Protestantism but by a secular scientific viewpoint. The technological achievements of Europe and America were unequaled elsewhere, and no agrarian army could withstand their mechanized military forces.

All this was the result of the industrial transformation, the only change in human history that compares in importance to the agrarian transformation. But unlike agrarianism, which had long since spread through the entire population, the major industrialized regions were still limited to Europe, America, and Japan. In less than 300 years, these areas had become separated from the agrarian ones – and not only separate, but in control. This division remains the most important one in the world today. How did it happen?

Precursors of industrialization

There is much speculation about the causes of the industrial transformation (e.g., Hobsbawm 1969; Pomeranz 2001; Diamond 2005), but we still have no definitive explanation. We can, however, point to important precursors that suggest why industrialization began in Britain during the eighteenth and nineteenth centuries.

Western historians speak of the "Middle Ages" as those depressed centuries between the glories of Imperial Rome and the re-emergence (*renaissance*) of art and learning in fourteenth-century Italy. There were no synchronous centuries-long depressions in China or the Muslim world, so one might think they had a head start on laggardly Europe, though if they did, it came to nothing. Actually, there was no "Europe" before the Middle Ages. Ancient maps of the territories controlled by Alexander or by Rome give no indication of Europe as a separate entity. A continent is usually regarded as a large land mass surrounded by water. By that definition, "Europe" is nothing more than a large peninsular of Eurasia. Its inclusion among the world's continents depends on a unique land boundary drawn during the Middle Ages between Christendom and Islam. This boundary was easily crossed, most profitably by Venetian ship owners. Whether carrying Christian crusaders to fight for Jerusalem, or Oriental cargos of spices and silks to sell in Europe, there was plenty of money to be made by the carriers. This was the main source of wealth enabling the Italian Renaissance.

In 1453 Muslims captured the Christian city of Constantinople, which had long been a conduit for the east-west trade. Renaming the city Istanbul, its Turkish masters stopped the Venetians' lucrative sea trade, causing the economic decline of Italy. Most importantly, the blockade on Mediterranean shipping opened the prospect of enormous wealth for any European who could find a new sea route to Asia.

Voyages of discovery

With the Mediterranean route blocked, there were two options to reach Asia by sea from Europe's Atlantic coast: sail south down the West African coast an unknown distance, then, if possible, turn eastward toward Asia; or sail directly west across the Atlantic, circling the globe until reaching Asia, which was Columbus's plan. Educated people of the time knew that the world was round and that in theory one could reach the East by sailing west. Columbus's opponents disagreed with him primarily on the distance that would have to be sailed to reach Asia, which he put at 2,500 miles and they put at four times that distance. They were closer to the truth (about 13,000 miles), but no one knew that two new continents lay along the route at about the distance where Columbus expected to find Asia. Sponsored by Spain, Columbus ultimately made four voyages to America (1492–1504), always insisting that he had reached the Orient, although in the end hardly anyone believed him.

At nearly the same time (1497–1499), Vasco da Gama succeeded with the southward route, sailing around the Cape of Good Hope into the Indian Ocean and returning to Portugal with a valuable cargo of spices from India. Da Gama went again to India with a military squadron, killing and looting to establish by force a Portuguese commercial empire throughout the Indian Ocean, with trading outposts eventually reaching to China and Japan.

The sea routes to Asia and the Americas that Europeans discovered in the fifteenth and sixteenth centuries brought huge wealth to the "continent" that was invested in new trade and a general growth of commercial activity. They led to a vast overseas system of colonies that eventually supplied raw material for European factories and also the markets for finished goods. Shamelessly, the Europeans exploited the colonized areas for profit, personal glory, and spreading Christianity. They brought diseases to the New World for which Native Americans had no immunity, causing massive depopulation, which eventually led to the importation of African slaves to replace the diminishing number of Indian slaves. The leaders in overseas exploration, Portugal and Spain, spent much of their new wealth on warfare and conspicuous consumption, and ironically were among the last nations of Europe to industrialize. When the industrial transformation did begin in Britain, it quickly diffused through northern and central Europe and America – all Christian lands – but hit a wall at the cultural boundary of Islam. To this day, northern Africa and eastern Asia (now called the "Middle East") remain unindustrialized regions of the Third World.

Why England?

Details of European politics need not concern us, but it is worth noting the increasing stature of England, a beneficiary of the new Atlantic commerce. The island nation had naturally developed as a sea power, but it lacked colonies; its first New World profits came from raiding Spanish treasure ships returning from America. This and other factors led Spain to send a mighty naval armada against England in 1588, but it was utterly destroyed by a combination of bad luck, bad planning, and superior English seamanship.

Soon England (like France and the Netherlands) began seeking its own footholds in India and America. English colonies in North America did not produce gold and silver, like those of Spain, but did provide plenty of land for agricultural products that could be sent home. English colonists in America were themselves different than those from Spain, for they had come to stay, often as religious pilgrims, forming permanent and growing communities. By the middle of the eighteenth century, there were a million people of European descent in British North America, compared to 50,000 in the French sector and only 5,000 in the Spanish region. (The United Kingdom of Great Britain formed in 1707 when England and Scotland merged.)

The perennial European wars were temporarily settled in Britain's favor in the mid-eighteenth century. As a result, Britain gained full control of the eastern third of North America. At the same time, on the other side of the world, Britain became dominant in India and emerged as the world's mightiest colonial power, a position it retained for nearly two centuries, when it was literally true that the sun never set on the British Empire.

Britain's commercial role had been growing, making London the most populous city of Europe. Now its unique colonial situation, dominating both India and North

America with a mighty navy and merchant fleet, was a major factor leading to incipient industrialization in the decades after 1750. The colonies served two potential functions. They were huge sources of agricultural raw material, and their large or growing populations offered huge markets for manufactured products. All that was needed for a very rapid expansion of trade and profit was a technology that could inexpensively convert the raw material into finished goods.

The cotton textile industry was ideally suited to take advantage of this situation. Raw cotton could be shipped from the colonies to Britain where it was woven into cloth, some to be sold at home and in continental Europe, but much of it shipped to America and India for sale. Manufacturing capacity could be increased rapidly with relatively little capital, producing very high profits for a modest investment, as long as there was a market for the finished cloth, which there was. British textile manufacturers were not a necessary link in this chain since cotton grown in the colonies might have been woven there, but British policies discouraged colonial manufacture, with the result that the English became essential middlemen.

Producing cotton textile first requires spinning cotton tufts into thread and then weaving the thread into cloth. Spinning was by far the slower of these processes and the bottleneck to increased production. With money to be made, there was an acceleration of improvements in technique, many of them very clever but none requiring principles that were strange to mechanics of the time.

Traditionally, spinners worked in their own homes, in the country or in small villages. Among the most important new devices to improve the efficiency of spinning were the "water frame" and the "mule," both driven by water power and too expensive and elaborate for a cottage spinner. As a result, entrepreneurs soon went to towns where there were supplies of unskilled workers and set up factories – actually large spinning mills – with several of these machines housed together and run by a common source of power, first water and then steam. More hand weavers were required to turn the plentiful thread into cloth. But soon the weaving itself was mechanized, eliminating the need for highly skilled craftspeople. Often they were replaced in the factory by unskilled women and children who tended the mechanical looms.

During the first half of the nineteenth century, five- or six-storied factories, each with a tall chimney exhaling black smoke, became common in manufacturing cities such as Manchester. The factory owners employed urban workers who were wholly dependent on the low cash wages they earned, having no farm or cottage craft to fall back on. Factory work was specialized, repetitive, and tedious, usually requiring little skill but long hours, unlike traditional crafts and trades that varied by time of day and season and where one moved through a series of tasks in order to accomplish the worker's goals. Also, in preindustrial times there was a personal relationship between master and servant, or lord and serf, which implied rights and obligations on both sides, if unequal ones. The relationship between factory worker and employer was wholly impersonal, based solely on the exchange of wages for hours worked, a relationship that could be severed at short notice, with the employer having no responsibility for the welfare of former employees.

Factory work was not all bad, for even the workers themselves eventually benefitted from the prosperity produced by efficient mass production. Others came out much better, not only the factory owners but merchants, financiers, shippers, and traders who prospered from the rapid acceleration of business. Some at the top of the heap entered the ruling class, such as the businessmen Peel and Gladstone, whose sons became prime ministers. Many more moved into the new "middle class" – below the aristocracy but above the peasants and laborers – with comfortable homes far removed from the slums. Whereas the primary cleavage of the agrarian system had been between the hereditary aristocracy and the peasantry, that of the industrial system was between those who had acquired wealth, or at least a comfortable living, and those who lived in or near poverty.

After losing the War of American Independence, Britain could no longer prevent the establishment of textile factories in New England. Now the raw cotton from plantations in the South could be spun, woven, and brought to market without ever crossing an ocean. Industrialization in America soon rivaled that of Britain. The commercial nations of northern Europe joined in, building their own factories and competing for the markets that had hitherto been the exclusive domain of the British. With so many nations selling textiles, demand was saturated, and opportunities lessened for high profits from rapid expansion. The first growth stage of industrialism, based on cotton textiles, had reached its limit by the mid-nineteenth century when the second stage – based on coal, iron, and steel – began its spectacular takeoff.

Opening the age of fossil fuel

The expanding cities required fuel, and since wood was relatively scarce in England, there was increased demand for coal, which was plentiful. Many of the deeper coal mines had water seepage, but with heightened demand driving up prices, it was worthwhile draining the mines. In 1712, Thomas Newcomen installed a primitive steam engine for this purpose. The engine was a large cylinder containing a piston. Steam, produced by boiling water with a coal fire, entered the cylinder through a value, driving the piston upward. Then cold water was squirted into the cylinder to condense the steam, causing a vacuum so that atmospheric pressure drove the piston back down into the cylinder. By connecting the piston to a water pump, the up-and-down motion was harnessed to lift water from the mine, and these engines were used widely in the coalfields. See http://en.wikipedia.org/wiki/Newcomen_steam_engine for an animation of the Newcomen engine in operation.

James Watt is often credited with inventing the steam engine, but what he actually did was improve Newcomen's engine. Most importantly, Watt recognized how wasteful it was to squirt cold water into the cylinder to condense the steam, for it had to be reheated for the next stroke. His great improvement was to connect a separate condenser vessel to the cylinder, with a valve between them. In Watt's engine, like Newcomen's, steam was let into the cylinder, forcing the piston upward. But then the steam was allowed to expand from the cylinder to a separate cold condenser (by opening a valve between them), and it was there that the cooling occurred, producing a vacuum that allowed the piston to fall. By this device, the cylinder stayed hot for each stroke and required no reheating. Watt's engine was much more efficient, and it could stroke rapidly. Soon it was replacing water power in the textile factories. By 1807, the American Robert Fulton had put one on a boat to turn the paddle wheel, and lots of people began experimenting with steam locomotives that would run on tracks.

In 1825, it made good economic sense to construct a railroad to haul coal about ten miles from inland mines to a port on the English coast. By 1850, another 6,000 miles of railroads were opened in Britain, many of them not economically worthwhile in a country where most points were within easy access to water transport by sea, river, or canal. Nonetheless, investors who had accumulated capital from the textile industry, which now had limited opportunities for further growth, poured their money into railroads. In other ages, they might have spent their surplus wealth on displays of conspicuous consumption such as grand palaces, but now the modern capitalist spirit encouraged the investment of surplus money to make even more money. Railroads were especially sensible for crossing vast distances of land that lacked easy water routes, such as the east-west route across the United States, for which continuous track was completed in 1869.

The growth of railroads was astounding, whether for good economic reasons or because of investment mania. In 1830, there were less than 100 miles of track in the world; by 1850, there were over 23,000 miles of track. By 1900, there were 200,000 miles of track in the United States alone. The expanding railroad demanded more coal, iron, steel, and engines. Each of these became the basis of its own heavy industry, attracting more investment, which was accompanied by technological improvements on all fronts. By the 1880s, steel steamships eased ocean travel, initiating the massive wave of "new immigration" to the United States. Each industry fed on the others: steel machines mined and transported coal and iron which were used to produce steel, which was used to make machines, and on, and on. By this time the British population of Australia – originally criminals exported to the penal colony, later immigrants seeking new opportunities – began its own industrialization.

Why Japan?

With merchant and naval fleets of steel steamships, the newly constructed Suez Canal (1869) allowing Europeans direct access to the Indian Ocean, and railroads crossing territories heretofore accessible only with great difficulty, the industrial powers extended their presence to most other areas of the world. Some non-industrialized countries welcomed this presence, most had little to say about it, and some actively resisted. The Chinese wanted nothing of "barbarian" goods or culture from Europe, but they would trade their silks, tea, and porcelain for opium, which the British brought from India. When Chinese authorities tried to stop the drug trade, Britain's gunboats fought two Opium Wars (1839–42, 1857–60) to enforce its trading rights. The Japanese, too, wanted to bar Western trade; this time, American steam-powered gunboats under Commodore Perry opened the way (1853–54). These two Asian powers responded differently to the humiliation of Western domination. To the extent that they could, the Chinese closed their eyes to it.

The Japanese elite, impressed by American military might, decided that the best way to avoid domination was to achieve such power themselves. Over the next half century, a period called the "Meiji Restoration," Japan imported industrialization and its institutions from the West with remarkable speed. A new army with Western uniforms and armaments used commoners as soldiers rather than traditional samurai warriors. When some of the samurai revolted against Meiji changes, they were defeated by the army, thus verifying the superiority of modern methods over traditional warfare. During the Meiji reforms, the Japanese were guaranteed civil liberties, literacy was increased through compulsory education, and peasants were given legal ownership of their lands. Railroads, telegraph lines, steamships, central banking, postal systems, and industrial methods were copied from Europe and America. British and French textile processes were adapted to silk cloth making. Among the imports from the United States, baseball became a Japanese passion.

Soon Japan was joining Europe and America in a third stage of industrialization based on electrical devices and vehicles with internal combustion engines fueled by petroleum, including automobiles, trucks, tractors, ships, and airplanes. Thus, by the early twentieth century, two of the pillars of the Fossil Fuel Age, coal and oil, were in place. The third, natural gas, being harder to transport from wellheads to end users, did not come into its own until reliable long-distance pipelines were developed in the mid-twentieth century.

Japan also copied Western imperialism, beginning in the 1870s to forcibly "open" Korea and China, as Perry had opened Japan. A clash of expansionist interests with Russia led to the Russo-Japanese War in 1904–05 and Japan's heady victory over a major European power. Joining the Allies in World War I, Japan took over Germany's colonial possessions in Asia and the Pacific, emerging as the dominant power in the East. In the hard Depression years of the 1930s, Japan, like Germany, abandoned its democratic forms and adopted militaristic expansion as a national goal, bringing its interests in the Pacific head to head with the United States. The end result, of course, was the devastation of World War II.

Why not China?

If our extraterrestrials, making a reconnaissance in 1500, had been forced to report which of the earth's agrarian civilizations looked most impressive, they might well have chosen China. Even earthlings of that time knew that China (Cathay) was the source of the most desirable manufactured goods, perhaps the richest dominion of the day. Among the better known inventions that Europe borrowed from China are papermaking, printing, the compass, gunpowder, and cast iron. What was not known in Europe, but if recognized would have impressed even more, was that 80 years before Columbus's voyages, Admiral Zeng He commanded Chinese fleets of 60 or more ships that crossed the Indian Ocean to visit the east coast of Africa. Zeng's own ship of nine masts was four times the length of Columbus's flagship, *Santa Maria*. If not for the Ming emperor forbidding more voyaging – his reasons are obscure but apparently were related to domestic priorities – China might have discovered the sea routes to Europe or the Americas, with barely imaginable consequences.

By the nineteenth century, the industrial transformation of Europe and America had given them enormous advantages in wealth and power. It was not that China faltered but that the industrial nations surged ahead. While Japan industrialized, the Chinese did not. China's passivity might have been a sustainable posture if the West's presence had been limited to the coastal trading enclaves conceded by the emperor. But American and European missionaries searched inland for souls and converted many of them. With trade and Christianity came steamboats and railroad tracks, telegraph lines, and mining equipment, all offending nature's spirits and eliminating Chinese jobs. These modernizations – some envied, some hated – undermined the traditional system. By 1895, the newly empowered Japan wrested Korea from China, a humiliating defeat for the Chinese, who had long regarded the Japanese as an inferior people. In 1911 Chinese revolutionaries overthrew the Qing dynasty and the last emperor, a little boy named Pu Yi.

Dynastic rule was replaced by a republic, though at no time from 1912 until the Japanese occupation of 1938 did the government function well as a parliamentary system. During and after World War II, China was in continual turmoil. Mao Zedong's Communists decisively won control in 1949, while his American-supported opponents, the Nationalist Chinese, fled to the large offshore island of Formosa. Renamed Taiwan, this island is today an effectively separate (and successfully industrialized) capitalist democracy although mainland China continues to regard it part of greater China.

The great divergence

Extraterrestrials circling the earth this evening would actually see the difference wrought by the industrial transformation. Figure 1.1 is a composite of NASA satellite images taken on cloudless nights. Today's industrialized regions – North America, Europe, and Japan – are brighter than the rest of the world, except for its largest cities.

If in the early twentieth century a list had been compiled of then-industrialized nations and another was compiled now, the two listings would not be very different. The "great divergence" that had occurred by World War I was sufficiently wide and enduring that few nations have since crossed the gap. The Soviet Union was a major exception.

FIGURE 1.1 The earth at night shows the electrical brilliance of North America, Europe and Japan (NASA)

In 1917, when the Russian Revolution ended the rule of tsars, and Lenin began the creation of the Union of Soviet Socialist Republics under the rule of the Communist Party, that new nation was technologically backward, much of its population barely removed from peasantry. Lenin and his successor Stalin wrought an industrial transformation at a terrible cost, including millions of lives, yet it worked for half a century. Hitler was broken at least as much by the Soviet industrial army and materiel suppliers as by the Allies attacking from the west. During the Cold War, the Soviets nearly equaled or surpassed America in military and space technology, in the development of energy and electrical resources, and in pure science and mathematics. The center and east of that immense nation were richly endowed with fuels and other resources devoured by industry, but the people who embodied scientific, technological, and artistic expertise were concentrated in the European part of the country, as were the reins of government. As a result, when the USSR collapsed, its 15 republics becoming 15 separate nations, those with populations concentrated in Europe, including rump Russia itself, retained their status as industrial (if temporarily impoverished) nations, while those in Central Asia essentially became again Third World countries.

By this time there had been a few more additions to the industrial list. Israel, created in 1948 as a refuge for survivors of the Holocaust, quickly developed using the skills of its European immigrants, partly to meet the need for a high tech defense against hostile neighbors, and considerably aided by financial and technological transfers from the United States. The industrialization of South Korea and Taiwan, modeled after Japan, was kick-started by the US crusade against Communism. South Korea benefitted from huge transfers of money and technology during the Korean War (1950–53) and the standoff with North Korea afterward. The US aided Taiwan to forestall attempts by mainland China to absorb the Nationalist Chinese who had fled to that island, and it received more American resources as a reserve area during the Vietnam War.

There has been unresolved debate, largely ideological, over the major reason that very few nations made an industrial transformation during the twentieth century. At one extreme is the argument that the industrial nations used their power to exploit and hold down colonial possessions. Even with the retreat of European colonialism after World War II, the economic and covert military might of the industrial powers, especially America and the Soviet Union, maintained this repression, preventing any real progress among Third World nations, excepting those newly rich because they sit on valuable pools of oil.

Probably no historian denies some truth to this argument, beginning with the preindustrial decimation of Native American populations through diseases brought to the New World by Europeans. Exploitation of Sub-Saharan Africa is epitomized by, but certainly not limited to, Belgium's King Leopold II taking personal possession of the entire Congo, doing what he pleased with its resources and people. Following World War I, the British and French agreed in secret to divide up the conquered Turkish Empire, drawing the boundaries for what would become many of the nations of today's Middle East and setting in place their initial rulers. I have already mentioned the British and American incursions into Asia. There is no debate about whether such offenses occurred, but about whether that exploitation is the primary reason that so few Third World nations industrialized during the past century.

At the other extreme is the argument that Third World nations were not held back so much as the industrial nations jumped ahead. Life in most of today's Third World is no worse, and in some ways better, than life was throughout the entire world just two or three centuries ago. To illustrate this position, I recommend an animation by Professor Hans Rosling of Sweden's Karolinska Institute, available on the internet at http://www.gapminder.org/videos/200-years-that-changed-the-world/. Using measures normally regarded as indicators of progress, life expectancy and gross domestic product (GDP) per capita, the animation shows the world's nations as very similar (in these terms) at the beginning of the nineteenth century, then diverging as the industrial nations left the rest of the world far behind, but not worse off than they were at the start.

Conclusion

From 10,000 until about 200 years ago, humans lived in agrarian societies, their energy sources renewable, and their power supplied by muscle, wind, water, and burning biomass. The industrial transformation of Europe, North America and Japan, based on new technology and fossil fuels, produced great wealth and power but also the full range of modern environmental problems, many of them associated with energy.

It is fairly easy to trace the historical trajectory of the industrial countries over the past few centuries into the Fossil Fuel Age, but difficult to understand the underlying causes of this epochal departure from the agrarian past. Heated ideological arguments claim to explain the great divergence that exists today. One side blames colonial exploitation by the industrial powers for preventing today's poorer nations from developing; the other side eludes blame, claiming that the industrial powers did not so much exploit and inhibit the rest of the world but leapt ahead of it.

It seems to me that overall, the great divergence reflects the remarkable "progress" made in nations that did industrialize. Which brings us to the crux of this book: How can the industrial nations continue to enjoy their relatively new benefits, and spread them to other nations, while minimizing the problems caused by their massive consumption of energy?

2

IS THE MALTHUSIAN TRAP IMMINENT? ARE OTHER PROBLEMS MORE WORRISOME?

The idea that a growing human population would outstrip its resources was popularized by the Reverend Thomas Malthus, writing in England in 1798 at the beginning of industrialization. Malthus was struck by the country's rapid population growth, far faster than any foreseeable increase in agricultural production. He warned of an inevitable calamity when the number of people surpassed the food supply, bringing an era of persistent famine and suffering. To slow the growth of population, the minister advocated sexual abstinence, a solution as unrealistic in his day as in our own.

A "neo-Malthusian" version of the global problem of population and resources is widely accepted today. It holds that the world is trapped between a growing population and finite resources, with the inevitable outcome being severe shortages and widespread suffering. Therefore it is imperative for families and nations to limit new births, and to stop using energy from fuels that are depleting, turning instead to sources that are renewable because they are derived from sunshine: solar energy, biomass, hydroelectricity, and wind power.

There is an undeniable logic to the Malthusian trap. Continually growing consumption of any finite resource must eventually cause depletion. In the US the argument is most often applied to petroleum, less to coal that is so abundant it would take centuries at current consumption rates to use up what is economically available.

"Easy oil," that which can be extracted relatively profitably with the technology of the time, is most vulnerable to depletion in the foreseeable future because giant underground reservoirs are fairly well mapped around the world, and it is generally assumed that there are not many more of these "elephants" to be discovered. Some known reserves are protected from development by environmental regulation, for example, in the ecologically sensitive Alaska National Wildlife Refuge, or close offshore from scenic or populated coastlines. Other known or suspected deposits are not accessible at costs or with technology that is today workable. Some large oil deposits have been effectively pumped out in the sense of no longer bringing sufficient returns on investment. Output from some currently active fields is diminishing, and eventually these too will be abandoned as unprofitable. The United States, blessed with some of the largest known oil deposits, had its peak years of production around 1970, with domestic output subsequently waning (until 2005, when it increased again).

Why don't we run out?

The puzzle is that over more than two centuries since Malthus warned of persistent famine, and others have warned that we will run out of oil or other finite commodities, it hasn't happened. The paradox is well illustrated by a famous bet in 1980 between the neo-Malthusian Paul Ehrlich, an ecologist at Stanford University, and the anti-Malthusian Julian Simon, an economist at the University of Maryland (Tierney 1990). Simon argued that the more people the better because there would be more minds to come up with more ideas to make the world a better place. He proposed a bet to Ehrlich, as follows. Ehrlich could select any natural resource and any future date. If the resource became scarcer as the population grew, then its inflation-adjusted price should rise. Simon wanted to bet that due to human ingenuity the prices would fall. Ehrlich and two colleagues put $1,000 on five metals (tin, copper, nickel, chromium, and tungsten) and selected 1990 as the end date. Simon agreed to the terms, and a contract was drawn.

From 1980 to 1990, the period of the bet, while world population grew by more than 800 million, inflation-adjusted prices of all five commodities fell. Simon won. Ehrlich offered ad hoc reasons why there had not been price-raising scarcities of those particular commodities in that particular decade, insisting that they would become more costly in the future, and he may have been right. Figure 2.1 shows trends in the metal prices from 1950 to 2010 (in constant 1998 dollars per metric tonne) complied by the US Geological Survey, and we can see that if the bet had been placed the prior decade, 1970–80, Ehrlich would have won. Still, as Simon would point out, the general picture is one of wild short-term swings but no general rise in metal prices, and that remains true if we look all the way back to 1900, the earliest year for which comparable data are available (http://minerals.usgs.gov/ds/2005/140/).

The resounding effect of the bet's famous outcome was a serious questioning of the Malthusian trap. As it was re-examined, there appeared good reasons why the finiteness of a resource was not as serious a limit as it had seemed, and some escape routes became evident. As the price of a metal rose, it gave incentive to find and extract more ore, or to use it more efficiently, or to find cheaper substitutes. "Tin cans" are now made of thin aluminum, and cans themselves have been largely replaced by plastic bottles. Plumbing contractors stopped installing copper pipes, using PVC pipes instead, and optical fibers made from silicon now

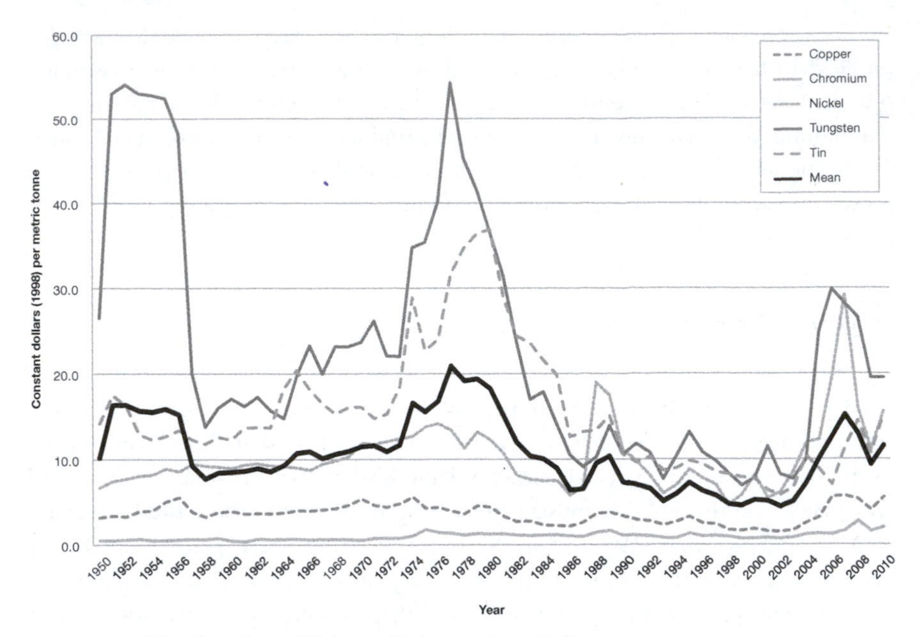

FIGURE 2.1 Yearly price of five metals in constant dollars per tonne (http://minerals.usgs.gov/ds/2005/140/)

serve functions once filled by copper wiring. Consider how these motivations and innovations apply to the finite supply of oil.

There is a popular image of oil being concentrated in large pools, many of them around the Persian Gulf, especially in Saudi Arabia. Pumping the oil from these pools is visualized as akin to sipping Coke from a glass, eventually draining the last drop. That image suggests we can entirely use up the world's oil, and it carries the corollary lesson that undiminishing (i.e., renewable) sources of energy we get from the sun are by that fact alone superior to finite fossil fuels.

In fact, like most finite resources, oil is distributed throughout the globe in many different ways. Lots of it does indeed sit in huge "pools" (actually oil-soaked sediments) around the Persian Gulf, easily and inexpensively tapped. Elsewhere it is deeper, under hard rock, and drilling is difficult. Much is buried deep beneath ocean water. Much oil is thinly distributed in shale rock. Canada has huge deposits of petroleum in Alberta's gritty tar sands. Abundant coal is another potential source because it can be chemically converted to oil.

If we include all of these sources in our inventory, there is a vast amount of oil in the earth, but most of it, at any given time, is too costly or troublesome to extract (Hall & Klitgaard 2011). As the easy oil runs out, people will still be able to get oil if they want it badly enough to pay the costs, develop the technologies, and accept the human and environmental hazards of obtaining it.

During the 1960s the oil-addicted United States began importing much of its petroleum from nondemocratic monarchies of the Middle East and elsewhere in the Third World. The reason was not that American oil was used up. Texas oil fields, once booming but later abandoned, still contained plenty of petroleum. But decades of pumping had reduced the pressure in underground reservoirs, making it increasingly difficult and expensive to pump up what remained. It was cheaper and more profitable for American oil companies to buy petroleum from the Persian Gulf, where it could be brought to the surface easily and inexpensively, and then shipped to America on supertankers. US oil production peaked in 1970 because it was cheaper and technologically easier to get petroleum overseas than to drill for it at home.

A fair-minded extraterrestrial might conclude that until the early 1970s, the industrial nations exploited the oil-producing nations of the Third World, paying them a pittance for their one valuable resource. The elites of these Third World oil producers, enriched by their overseas sales, knew that their petroleum was selling at bargain basement prices. Hoping to shift the balance, they banded together as the Organization of Oil Exporting Countries (OPEC), at first impotent against the industrial importers. Their opportunity to break the West's control of oil prices came in 1973 when the Yom Kippur War broke out between Israel, on one side, and the Arab nations Syria and Egypt on the other. Several OPEC nations are Arab, most importantly Saudi Arabia, and they rallied all OPEC members to boycott shipments of oil to nations that supported the Israelis in the war. The US, a staunch supporter of Israel, thus suffered an abrupt cutoff of oil shipments in 1973, igniting the first "energy crisis" of that decade. American consumers, used to a plentiful supply of cheap petroleum products, were suddenly parked in long lines at gas stations, and they were paying more to fill up their cars because the crimp in the overseas supply line immediately jacked up gasoline prices. A second oil shock occurred in 1979 when the authoritarian shah (king) of Iran, supported in power by the United States, was overthrown by Islamist revolutionaries. The upheaval in Iran, one of the major oil nations on the Persian Gulf, abruptly cut the quantity on the world market, raising prices of petroleum and its refined products, gasoline and diesel fuel, even more than in 1973.

With prices so high, consumers cut back on unnecessary use of petroleum by reducing their road travel, switching to more fuel-efficient cars, lowering thermostats, and converting from oil to natural gas for space heating. Utility companies switched from petroleum to other fuels for generation of electricity, and governments regulated or at least encouraged lower consumption and more efficiency in energy use. The rise in oil prices during the 1970s was one cause of the "stagflation" (an unusual combination of economic slowdown with rising inflation) during the last half of that decade, a further cause of lower energy consumption. In the meantime, with a barrel of petroleum commanding a higher price, oil producers in industrial nations found it worthwhile reopening previously "depleted" reservoirs and exploiting new production under the sea, which had before been too expensive and technologically challenging. Thus, the high prices of the 1970s produced a reduction in demand for oil, and an increase in production

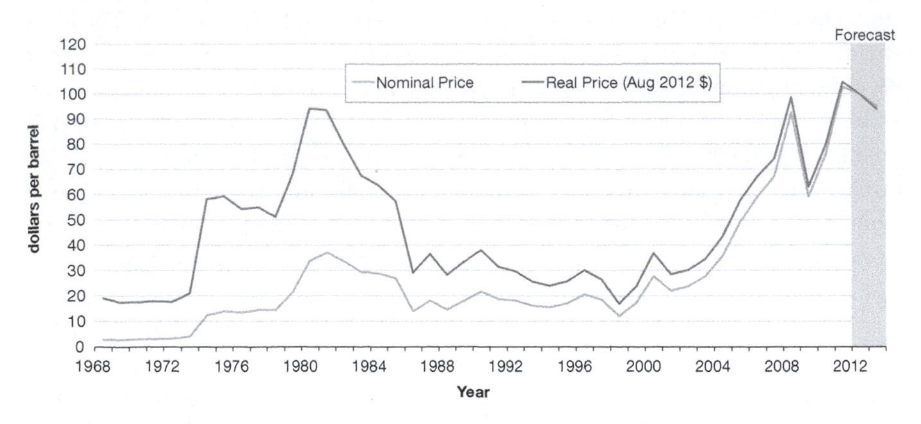

FIGURE 2.2 Annual crude oil prices (EIA)

of oil by the industrial nations themselves, a combination that cut the demand for OPEC oil.

By 1982, the price for a barrel of oil was rapidly dropping on the world market. To maintain the large cash inflows to which the OPEC nations had become accustomed, some OPEC members cheated on their agreed-upon export quotas, pumping more than their limits and effectively flooding the market with oil and further depressing prices. By the mid-1980s, a barrel of oil was relatively cheap again and the "energy crisis" a bad memory. No longer chafing under high gasoline prices, Western consumers and governments returned to many of their wasteful ways. This was the era when fuel-inefficient SUVs became popular with Americans and profitable for automobile manufacturers. Demand for oil turned upward again and with it the price of gasoline.

In the first decade of this century, with oil prices rising again, there was renewed incentive for fuel-efficient automobiles. In the United States, SUVs accounted for 18 percent of new-car sales in 2002, but only 7 percent in 2010. American oil producers reopened "spent" oil fields in Texas and elsewhere, using improved techniques to pump up more of the remnant deposits (Krauss & Lipton 2012).

This example of oil is informative about the Malthusian trap and the means of escape. Of course finite resources are ultimately limited by the amounts that are in the earth, but for practical purposes, supply is limited even more by the price that consumers are willing to pay, the extractive technologies that are available, and the degree of environmental damage, risks to health, and of accidents that people and nations are willing to accept. As the price of a resource rises, extractors will "dig deeper" to bring out what was previously not worthwhile; engineers will improve techniques for extraction. Sometimes it is affordable to mitigate health, environmental and accident risk by applying protections that were previously too expensive or technologically advanced to implement, but sometimes higher risks are simply accepted.

Malthus's original concern was the food supply, seemingly limited by the amount of arable land. He had no idea that agriculture itself would become industrialized and as a result far more efficient, with each acre of land producing more food, with less labor, through the use of tractors, combines, hybrid seeds, irrigation, and manufactured fertilizers and pesticides. These were the basis of the "Green Revolution" that since World War II has increased world supply of food faster than population has grown. Cows, pigs, and chickens are now produced in feedlots, efficiently and in quantity (if inhumanely), as if they were cars manufactured on an assembly line.

Famine is less a problem today than ever before, and when starvation does occur, as in Somalia or Darfur, it is usually not because of a lack of food but the inability to transport food aid from the dock across the lines of a civil war, or by the inability of poor people to buy food that is available in the markets. These tragedies are better described as sociological problems than as Malthusian limits.

With birth rates declining nearly everywhere except Sub-Saharan Africa, the United Nations projects that global population will level off by the end of the century at about ten billion people, compared to between six and seven billion today (www.un.org/popin/wdtrends.htm). Imagine that the quantity of fossil fuels available for these people was nearly inexhaustible. There would be no Malthusian trap in that imaginary world, at least with regard to energy. Does it follow that there would be no serious problems caused by very high consumption of energy and electricity? No, there would still be problems aplenty, and we are suffering them already. It is worth a quick listing of non-Malthusian problems, those that would be troublesome even without any global depletion of fuels.

Climate change

Global warming may or may not be the most serious problem associated with consumption of fossil fuels but in either case is remarkable for the number of Americans, particularly conservative Republicans, who do not believe it is occurring, or worse, believe it a myth perpetrated by conspiratorial scientists to improve their access to federal research funding (McCright & Dunlap 2011; also see Chapter 11).

The reason for global warming is easy enough to understand and not controversial. Sunlight is electromagnetic radiation, traveling at the speed of light, reaching earth in a little over eight minutes. When it encounters molecules in our atmosphere, some sun rays are reflected back out to space, and some continue through to the surface of the earth. A ray of sunlight impacting the earth gives up some of its energy in the form of heat. That ray, now with lower energy, would also reflect back into space unless in passing again through the atmosphere it runs into a CO_2 (carbon dioxide) molecule. It has been known for over a century that CO_2 molecules act like one-way gates, allowing incoming high-energy sun rays to pass, but blocking outgoing lower-energy rays from leaving. If blocked, the ray hits the earth again, adding still more heat. The more CO_2 there is in the atmosphere, the more sun rays will be blocked from leaving. The result is a warming of the earth's land, sea, and atmosphere. There are other molecules that act like one-way gates, including water vapor and methane, collectively called "greenhouse gases" by analogy to the warming effect of a glass greenhouse.

Earth's atmosphere contains only a tiny portion of CO_2, less than four-hundredths of one percent, most of that present well before the appearance of humans. If not for the greenhouse gases naturally in the atmosphere, earth would be too cold to support life. There are other natural determinants of average global temperature including long-term variation in the earth's orbit around the sun, which seems to be the major cause of ice ages and intervening warm periods. The problem of concern is the change that is occurring since the onset of industrialization, as burning of fossil fuels increases the proportion of CO_2 in the air.

What most of us call "burning," chemists call "oxidation," the chemical combination of oxygen with another molecule. When the other molecule is a hydrocarbon, the chemical reaction produces heat along with new molecules, one of them carbon dioxide (CO_2). When oxidation occurs rapidly, as when burning logs in a fireplace, the heat appears as flame, hot to the touch. If the oxidation is slow, as in the decomposition of a dead log in the forest, the heat output is barely discernible. Except for the speed of reaction, the process of oxidation is the same, giving heat and CO_2 among its end products.

Charles Keeling began making direct measurements of atmospheric CO_2 in 1958 at the Mauna Loa Observatory in Hawaii, far from pollution sources. His results, now called the "Keeling Curve," show a continuing increase in atmospheric CO_2 from about 320 parts per million (ppm) in 1958 to roughly 390 ppm in 2012. Preindustrial levels are estimated at about 280 ppm, so there was about 40 percent increase by 2012. A doubling of the preindustrial level, which would make atmospheric CO_2 higher than it has been for tens of millions of years, may occur by the middle of this century, with China's yearly contribution now exceeding that of the United States, and India coming up quickly.

Much scientific debate during the 1990s focused on the adequacy of attempts to measure changes in average atmospheric temperature over the past century or more. There have been continuous thermometer records for that long, mostly from weather stations on land in the Northern Hemisphere. With the subsequent increase in number of stations, and their more widespread distribution, there were uncertainties in how to calculate average global temperatures that were properly comparable from one decade to another. Some early weather stations established in the countryside were enveloped by expanding urban areas, and since urban areas are generally warmer than rural areas (due to the concentration of combustion sources and heat retaining properties of asphalt, cement, and other building materials), there was concern that apparently rising temperature was actually an artifact of cities encroaching on the thermometers. The increased pace of research in the past quarter century has mostly solved these problems. Yearly calculations of average global temperature are now widely accepted as showing about a

0.6° C (1° F) rise since the late nineteenth century, with recent decades the warmest on record.

Furthermore, analysis of proxy measures for temperature that extend much farther back in time, including tree ring and coral growth, and composition of lake sediments and of glacial ice cores, show the present climate to be the warmest in the last millennium, if not longer. A doubling of preindustrial CO_2 will produce a temperature increase that is uncertain and controversial, but credible estimates range from 1.5 to 4.5° C (about 3 to 8° F), sufficient to cause a serious rise in sea level and other changes in the earth's environment.

Air and water pollution

On the first Earth Day, April 22, 1970, no one worried about global warming, which would not become a public issue until 1988. The major concerns in 1970 were pesticides (especially DDT), and air and water pollution, which had reached critical levels in major industrial cities. London, used to heavy fogs mixed with coal smoke, had a particularly bad "killer fog" in 1952, causing an estimated 4,000 premature deaths. The mixture of smoke and fog gave rise to hybrid words such as "smaze" or eventually "smog."

As coal burning diminished in industrial cities, due partly to new antipollution regulations, a different kind of smog became the chief problem of urban air, the product less of coal smoke than of exhaust fumes from cars and trucks, particularly potent in sunny regions. Flying into Los Angeles during the postwar decades sometimes meant descending into an opaque brownish cloud. The city had smog alerts on days when the air was so bad that schools did not let children outside for recess.

Pollution of water was as bad as of air. Historically the worst pollutant of lakes and rivers had been raw sewage, largely controlled by the early twentieth century, but in 1970 it was industrial waste. The Cuyahoga River running through Cleveland, Ohio had so much oil floating atop the water that it would occasionally catch fire. In 1969 a blowout at a drilling rig off the shore of prosperous Santa Barbara, California triggered a large release of oil that tarred beaches and killed wildlife. Angry citizens, with the help of journalists, woke the nation to the threat of water pollution.

With considerable unanimity, certainly compared to controversies at that time over Vietnam and civil rights, the US Congress created a new federal agency to control pollution, the Environmental Protection Agency (EPA), and passed with bipartisan support the Clean Air Act of 1970, signed into law by Republican President Richard Nixon. Soon to follow was a Clean Water Act. In 1990, the most important strengthening of the Clean Air Act was supported by President George H. W. Bush, another Republican, who labeled himself "the environmental president."

The Clean Air Act required EPA to set standards for six common air pollutants: ground-level ozone, particulate matter (i.e., very small particles), carbon monoxide, nitrogen oxides, sulfur dioxide, and lead. (Carbon dioxide was not then defined as a pollutant.) Lead in air was most dramatically and easily reduced because most of it came from tetraethyl lead that was for decades added to gasoline to eliminate engine "knocking." By banning lead in gas, governments quickly eliminated most of the lead in air. Requiring catalytic converters in the exhaust pipes of internal combustion engines reduced emissions of carbon monoxide, nitrogen oxides, and unburned hydrocarbons. Mandated reductions in sulfur dioxide came partly from fuel switching: coal and oil with high sulfur content was replaced with low-sulfur coal and oil. Also "scrubbers" installed in the smokestacks of coal- and oil-burning power plants removed sulfur dioxide from stack gases. Reductions in ground-level ozone and particulates have been less successful, but US levels have not worsened despite considerably increased fuel usage.

The new antipollution laws were so successful that similar measures were enacted across the industrial nations. While air and water of the high energy-consuming industrial world became cleaner, pollution in the Third World generally worsened and today is awful (Timmons & Vyawahare 2012). During a "bad air" day in Beijing, one can barely see across the street, and this is not simply a matter of aesthetics; pollution sickens and kills. The World Health Organization (WHO) estimates that inhalation of very small particles in air pollution, those under ten micrometers in diameter, cause over a million deaths annually, disproportionally in the Third World. But industrial nations cannot rest on their laurels because their low levels of pollution (compared to the Third World) still damage health and the environment, and the more fuels and materials that we consume, the more polluting waste we must deal with. Appendix I provides more detail on air pollution.

Health risks

Dramatic accidents in the industrial nations, widely reported in news media, become vivid emblems of the lives lost to energy production. In 1986 the Chernobyl explosion and meltdown in Soviet Ukraine – at first kept a state secret but later admitted – caused 30 confirmed deaths among reactor personnel and emergency workers, most due to acute radiation disease within a year of exposure. It is impossible to assess with any precision the numbers of fatal cancers that will eventually be caused by Chernobyl because these may take decades to appear and would not be discernible from the far larger number of cancers that occur naturally. By one estimate, among the 600,000 people receiving significant exposures, the possible increase in cancer mortality might be up to 4,000 cancer fatalities. Among the five million people residing in other contaminated areas where doses were much lower, the increase in cancer mortality is expected to be less than one percent. Thyroid cancer in children or adolescents is the fastest cancer to appear after irradiation. From 1992 to 2002 in Belarus, Russia, and Ukraine more than 4,000 cases of thyroid cancer were diagnosed among those who were 18 years old or younger at the time of the accident. Given the rarity of thyroid cancer in youth, it is likely that a large portion of these is attributable to the accident. Normally thyroid cancer is managed without a decrease in longevity; however, 15 deaths were related to progression of the disease (Chernobyl Forum 2006).

No immediate deaths are directly attributed to Japan's *Fukushima Dachii* nuclear accident in 2011 after a huge tsunami hit the nation's western coast. (The tsunami, seen around the world in horrific videos, caused about 22,000 people to be declared dead or missing.) At least six nuclear plant workers received radiation exposures above the legal lifetime limit. Some 80,000 people were evacuated from around the Fukushima complex. Preliminary estimate of long-term health effects to people offsite is minimal, though that may change as more data become available (ANS 2012). Perhaps the worst health effects of Chernobyl and *Fukushima Dachii* are the emotional stresses and stress-related diseases in families fearful of long-lingering radioactive contamination.

In 2010 the explosion at the Deepwater Horizon drilling platform in the Gulf of Mexico, causing the largest offshore oil spill in US history, killed 11 people. In 2010, 29 miners died from an underground explosion in a Massey Energy coal mine in West Virginia, the worst coal mining accident in the US since 1970, when 38 miners were killed in Kentucky.

These awful events have become enormously symbolic, but today energy-related accidents in the industrial nations are far less severe than they used to be, and they are relatively few and infrequent compared to energy extraction accidents in the Third World. Though rarely reported by Western news media, there are coal mining accidents nearly every week in China, killing 5,000 miners yearly, the worst rate in the world. In the United States, which mines about 40 percent as much coal as China, there are about 30 mining deaths per year (Barboza 2007).

Despite their drama and symbolism, accidents are relatively minor causes of death and illness from energy consumption. A much larger toll is caused by pollution from normally operating energy technologies, especially using coal and oil. Inhabitants of industrial cities breathe air that is far healthier than people in Chinese or Indian cities, but the cleaner air of developed nations is still troublesome. The US EPA warns that ground-level ozone even at relatively low levels can harm children, older adults, and people with lung disease or a particular sensitivity to ozone. Breathing ozone can trigger chest pain, throat irritation, and congestion; it can worsen bronchitis, emphysema, and asthma; it can make the lungs susceptible to infection, reduce lung function, and inflame the linings of the lungs, eventually scarring lung tissue.

WHO estimates that outdoor air pollution causes 1.3 million deaths worldwide per year, disproportionately in poor nations. In homes where biomass fuel or coal is used for cooking and heating, particulate levels may be 10–50 times higher than guideline values. Mortality in cities with high levels of pollution exceeds that observed in relatively cleaner cities by 15–20 percent.

Particulates in the air affect more people than any other pollutant, and the smallest particles ($PM_{2.5}$) are the worst because they penetrate more deeply into the lungs. Even in the European Union, according to WHO, average life expectancy is 8.6 months lower due to exposure to $PM_{2.5}$ from human activities (http://www.who.int/mediacentre/factsheets/fs313/en/index.html). The American Lung Association estimated in 2011 that if EPA placed very stringent limits on the smallest particulates, $PM_{2.5}$, each year the United States would be spared 35,700 premature deaths, 2,350 heart attacks, 23,290 visits to the hospital and emergency room, 29,800 cases of acute bronchitis, 1.4 million cases of aggravated asthma, and 2.7 million days of missed work or school (http://earthjustice.org/sites/default/files/SickOfSoot_2011.pdf). The specificity of these numbers veils uncertainties in the underlying research methodology, but few experts doubt that tiny particulates are hazardous even in the relatively clean-appearing urban air of industrial nations.

Environmental damage

There is no free lunch as far as energy is concerned. Every fuel or energy source causes environmental effects not wanted by someone – often not wanted by anyone. There is a common misperception that renewable sources of energy are an exception, environmentally benign: solar panels just sit there, making electricity from sunlight, not bothering anyone.

All renewable forms of energy come from sunlight. Windmills turn because the sun heats air unevenly, causing air currents to move from one place to another. Sunlight is required for the photosynthesis that makes biofuels. The sun evaporates water, which later falls as rain, filling rivers that flow to the sea, producing hydroelectricity.

An enormous amount of energy reaches the earth as sunlight. True, it comes only in daytime, and not as much on cloudy days, but there is a lot of it. However, it is very spread out, which is to say of low power per unit of area, so that a square meter of desert collects only a small amount of energy in an hour. One would need very many square meters of desert, or much wind turning windmills, to collect in one day the amount of energy that is in one liter of gasoline.

For all their disadvantages, the energy content of fossil fuels is very dense. They contain a lot of energy per unit weight or volume, far more than fuels like wood or corn or sugar cane that are being continually renewed by the sun. The traditional American fireplace is wood burning and very cheery, but it needs to be restacked with wood every half hour or so. It was more common in Britain to burn coal in the fireplace. Since there is much more energy in coal than in an equal weight of wood, the coal load burns much longer, giving much more heat, than the load of wood. (Now people are using gas fireplaces that can be turned on and off with a remote control. In America these are dressed with fake logs, in Britain with fake coal.)

So here is the crux of the environmental problem with renewables. You need a lot of area to collect the same amount of energy that is available in a small amount of fossil fuel. If that area is filled with solar panels or windmills, you must use a lot of material and energy to construct, transport, and install these, that is, before they are put in service. If you are growing crops to convert to ethanol or biodiesel, it takes a lot of cropland, and perhaps fertilizer and irrigation, as well as harvesting.

There are additional environmental effects, some specific to each source. Windmills, in practice, have been harmful to birds and bats, and they may be aesthetically objectionable if placed in scenic areas, or if placed anywhere in great numbers. Dams destroy wild rivers and interfere with fish trying to swim upstream to spawn; they create artificial lakes, flooding preexisting habitat; they silt up and sometimes fail catastrophically. Ethanol foolishly made from corn, a subsidized American policy, requires about as much energy input as is output to the gas tank; it requires farmland that might better be devoted to food production, and has elevated the price of corn on world markets, which is especially harmful to farmers in poor nations.

None of this is an argument against renewable energy. It is an argument against the naïve assumption that renewable energy is environmentally benign.

Resource wars

Humans, like animals, have always competed for desired but scarce resources, sometimes with lethal effect. A major innovation of agrarian civilization was the development of peaceful trade in resources, so that one could be exchanged for another to the mutual benefit of all parties. Another innovation of agrarian society was the ability to raise large armies with better (metal) weapons than existed in pre-agrarian times, so that when trade did not provide a satisfactory arrangement, and people resorted to force, the numbers slain could be far higher than in fights between small bands of hunters and gatherers. Ancient texts including the Bible claim that frightful numbers – even by today's standards – were killed in battles of the past, but these may be exaggerations, first because it would have been difficult to raise very large armies and keep them supplied in the field, especially when labor was needed during harvest season, and also because bronze and iron weapons, though better than stones and clubs, had limited efficiency. It required truly industrial warfare with large armies transported and supplied by trains and trucks, and armed with accurate guns, to produce the killing fields of modern times; for example, 50,000 killed, wounded, or missing in three days at Gettysburg (1863), or nearly half a million casualties in eight days at the First Battle of the Marne (1914).

There are many reasons for the threat or pursuit of war beside competition over resources, but this was an important one in the twentieth century. During the 1930s, Japan, largely barren of natural resources, sought to enlarge its access to oil and other material elsewhere in the Pacific basin, bringing it into competition with the United States. Hitler famously preached that Germany needed more living space (*Lebensraum*) and must expand to the east. His ultimately fatal decision to attack the Soviet Union was largely motivated by a desire to capture its oil fields on the Caspian Sea.

Since World War II the industrial nations have managed international affairs so as to avoid overt battle on their own soil (excepting the former Yugoslavia), instead fighting or promoting warfare in nations of the Third World. Many of these were motivated by Cold War ideology, with the US supporting one side and the USSR the other in numerous civil wars, sometimes trading sides, the essential strategy being to oppose whoever was supported by the other. So we may be skeptical of any claim that resource competition is the primary reason for war. However it surely promotes considerable expenditure on military preparedness even if no one is actually placed in firing position.

The clearest example of industrial warfare over oil was the Gulf War of 1991, waged by a UN-authorized coalition of nations led by US forces in response to Iraq's invasion of Kuwait. The year before, Saddam Hussein had rapidly attacked and annexed Kuwait, adding its substantial oil reserves to those of Iraq. This brought immediate sanctions against Iraq by members of the UN Security Council, partly because of the illegality of the invasion and annexation, but also for fear of having so much oil under the control of so despicable and manipulative a leader (Klare 2005).

Rather than withdrawing, Saddam moved his combat forces toward Saudi Arabia, where the largest proven oil deposits in the world were just over the border. Saudi King Fahd, seeing the obvious possibility of a further Iraqi advance, authorized the deployment of US forces to aid in his border defense, eventually hosting over 600,000 coalition troops. The US would spend $60 billion on the Gulf War, with the Saudis and Kuwaitis eventually covering half of that, a good investment for them.

The coalition attack began with an aerial bombardment in January 1991. This was followed five weeks later with an overwhelming assault by ground forces, liberating Kuwait and entering Iraqi territory within 100 hours, then declaring a ceasefire. During the fighting, between 20,000 and 35,000 Iraqis were killed against fewer than 400 coalition deaths.

The coalition of 34 nations that President George H. W. Bush assembled was so large – compare it to the small number recruited by his son for the Iraq war of 2003 – because few nations could tolerate Saddam Hussein controlling nearly all the oil on the western side of the Persian Gulf. There were strange bedfellows in the coalition: industrial democracies, OPEC members, other Third World nations. Israelis, acceding to an American request to *not* join the fight for fear of alienating Arab nations that were on board, sat the war out in air raid conditions, gritting their teeth as Iraq launched Scud missiles (inaccurately) toward the little country.

The true reasons for the second American-led invasion of Iraq, in 2003, remain puzzling. None of the rationales given by the administration of President George W. Bush, or by his critics, seem in retrospect sufficiently valid to provoke so drastic an action unless one assumes that the whole thing was a mistake. If one reason was the desire to maintain access to Iraqi oil, or to take control of it, the result was counterproductive because the war and its aftermath crippled the nation's ability to produce and export much petroleum.

Petrodollars, dictators, and terrorists

After the Gulf War coalition pushed Saddam Hussein's army back to Iraq, a TV comedian noted snidely that the US returned Kuwait to its rightful dictator. American military bases remained in Saudi Arabia with the king's approval to protect the stability and oil of the Saudi royals from any future threat by Iraq, and for continuing coalition missions to limit Saddam Hussein's reprisals against Iraqi opponents.

Nowhere is there a more blatant contrast to the ideals of Western democracy than on the Arabian Peninsula. King Abdel Aziz Ibn Saud unified the kingdom of Saudi Arabia through military conquests between 1902 and 1926. He built its legitimacy on a long-standing alliance between the Al Saud family and the Wahhabi sect of Islam. Since that time the royal family and the Wahhabi imams have supported each other in a mutually beneficial relationship. The Wahhabis preach a fundamentalist interpretation of Islamic law, opposing reforms that would bring Muslims closer to Western views about equality of the sexes, democratic rule, freedom of the press, separation of church from state, and elimination of archaic punishments such as stoning adulterers or cutting off the hands of thieves. Women tend to be sequestered and protected in Saudi Arabia. From a Western perspective, women have minimal rights, barely appear in the workplace, are required to dress conservatively, and cannot drive cars. Capital punishment includes beheading and stoning, which may be imposed for adultery, apostasy, and witchcraft. In brief, Saudi institutions and cultural values are as far from those of the West as can be found in the world today. But the Saudis have oil, more proven reserves than any other nation.

During World War II Saudi oil was undeveloped and therefore unimportant, but the United States and Britain recognized its potential and began to compete for the right to exploit it. In 1943 President Franklin Roosevelt declared impoverished Saudi Arabia vital to the defense of the United States, approving Lend-lead aid. In 1945 the ailing president, returning from his Yalta meeting with Churchill and Stalin, stopped near the Suez Canal for a shipboard meeting with King Ibn Saud, having previously written that he wished to talk about oil. The king arrived with a large retinue including the royal astrologer, sword-bearing attendants, and sheep to slaughter for meals. His party slept for two nights on carpets on deck rather than in staterooms, building charcoal fires to brew coffee.

The meeting was successful on a personal level. Roosevelt sat in his wheelchair, which the king, infirm himself, admired. Roosevelt had an extra wheelchair that he gave to his guest; it became a prized possession and a symbol of their friendship. The Americans also gave him a DC-3 airplane with a throne installed. (Shortly after, Churchill arrived for his own meeting with the king, gifting him a Rolls-Royce, but they did not establish the same rapport.) Records of the talks are sparse. One subject was Palestine; to Roosevelt's disappointment the king was disinclined to help European Jews settle there after the war.

Regarding oil, apparently there was some understanding, perhaps only implicit, about a reciprocal agreement that the United States could have exclusive rights to develop Saudi oil, and America would support Ibn Saud and his family (Yergin 1991; Bronson 2008). The king had already granted concessions for exploration to American oil companies, and these would merge into the Arabian American Oil Company, or Aramco. The Saudi government gradually acquired shares in Aramco, taking full control in 1980.

Also in 1980, amidst the energy crisis begun in the 1970s, US President Jimmy Carter declared in his state-of-the-union message, "Any attempt by an outside force to gain control of the Persian Gulf region will be regarded as an assault on the vital interests of the United States of America, and such an assault will be repelled by any means necessary, including military force." The doctrine was a response to the 1979 invasion of Afghanistan by the Soviet Union, intended to deter the Soviets from seeking control in the Gulf.

Saudi Arabia contains the cities of Mecca and Medina, Islam's most sacred sites, and is the Muslim Holy Land. Traditionally infidels were not allowed in, a prohibition clearly violated by the American bases remaining after the Gulf War. This was an affront to Islamists, made worse by the presence of female soldiers sometimes immodestly dressed. Among those gravely offended was Osama bin Laden, returned home from Afghanistan where he had led a contingent of Arabs in the war against the Soviets (ironically fighting on the side supported by the United States). He urged the king (unconvincingly) to send the Americans away, that he could oversee an Arab defense of the Holy Land from attack. This was one element of the schism that led to Osama's banishment from Saudi Arabia and eventually his move back to Afghanistan, where he planned the attack on America of September 11, 2001. A primary reason for targeting the United States was its insult to Islam from the US bases (Bergen 2001).

Osama was wealthy, his money inherited from his father, whose construction company, with close ties to the royal family, helped build Saudi Arabia's modern infrastructure. This was funded by money flowing in from the West to buy oil – "petrodollars." Despite Osama's estrangement from the king, he and his organization al Qaeda enjoyed financial and emotional support from the kingdom's Islamists. Fifteen of the 19 hijackers on September 11 were Saudi nationals. It is not a misstatement to say that the West largely funded with petrodollars the men who attacked it. The huge flow of petrodollars continues unabated, as do terrorist attempts to attack the West (Yetiv 2012).

Dollar/euro cost

It is difficult to say whether the dollar/euro cost of energy is a problem or a solution to the energy dilemma. Americans have generally paid less than Europeans for energy, partly because of an abundance of natural resources, partly because of generous government subsidies to energy producers, and partly because US state and federal taxes on fuel consumption are low. A $4 gallon of gasoline seems expensive to an American driver but cheap to visiting Europeans or Japanese.

Rapid escalation in fuel prices is hard on the pocketbook, reducing the ability to purchase other commodities. There are also indirect price hikes, for example, rising cost of aviation fuel passed on as higher airline ticket prices. There can be a recessionary effect on the economy, especially if the fuel being purchased is imported from elsewhere, sending more money out of the country.

On the other hand, cheap fuel encourages wasteful consumption. The joke goes that before the energy crisis of the 1970s, American cars used so much gasoline that you could never fill the tank if you left the engine idling. Only when gas prices at the pump rose in 1973 did Americans think seriously of fuel efficiency. When the cost of gasoline declined in the 1980s, Americans began buying fuel-inefficient SUVs.

A frequently proposed solution to excessive energy consumption and its associated problems is some version of increasing the cost of energy to discourage waste and reward conservation. Proposals come in various forms. A carbon tax, especially aimed at climate change, would increase the price of fuel according to how much carbon dioxide it added to the atmosphere. Though economically efficient, it carries the baggage of being called a "tax." Various "cap and trade" schemes would limit carbon and other pollutants while avoiding the term "tax," though in effect they too would raise the price of energy.

As important as the price level is the stability of the price. With energy prices lunging up and down from year to year, it is difficult for industries and consumers to make rational long-term plans. An investor in solar energy, for example, must be confident that his future sales will not be undercut by a sudden drop in the price of oil. These incentive systems are anathema to simple free-market fundamentalists, but the reality is that energy flow through an industrial system opens so many opportunities for manipulation that the notion of a wholly free market seems barely applicable.

Conclusion

Of course finite resources are ultimately limited by the amounts that are in the earth, but supply is further restricted by how much consumers are prepared to spend, the methods of extraction that are available, and the extent to which people and nations are prepared to accept environmental damage, risks to health, and accidents. As scarcity causes the price of oil or any other natural resource to rise, extractors will pay more to bring out what was previously not worthwhile, and engineers will improve techniques for extraction. There may be substitution of other resources, cheaper and more easily available, for the expensive resource.

One way or another, fuels will be available long into the future, so long as people and nations pay the price and bear the burdens of continually growing usage. These "costs" come in many forms beside cash, including climate change, pollution of air and water, environmental deterioration and destruction of habitat, accidents and ill health, wars over contested resources, and the transfer of huge streams of money to countries and private parties that undermine or actively attack the industrial nations that are financing them. In the foreseeable future, these may be more worrisome problems than the possibility that we will use up our finite resources.

3

ENERGY AND ELECTRICITY FLOW IN AN INDUSTRIAL SOCIETY

Energy and electricity are nearly always treated as topics of physical science or engineering. My intent here is to describe the highly complex flow of energy and electricity through an industrial society, from primary fuels to end uses, in a way that is meaningful and useful to social theorists and policymakers.

The role of energy in societal development has occasionally been incorporated into social theory (Cottrell 1955; White 1959; Lenski 1970) though in a way that today seems too simple. Despite occasional later contributions (Rosa et al. 1988; Schipper & Meyers 1992; Smil 1994), there is still no comprehensive body of research that can be called the "social theory of energy" (McKinnon 2007).

The flow of energy through industrial society

The simplest model of energy flowing through a population, common in writing of the 1960s and 1970s, presumes that the natural environment has a limited carrying capacity, able to support relatively few individuals. Diagramed in Figure 3.1, the area of the box represents the size of the population living "off the land." Inputs to the box are nature's resources: sunshine, precipitation, soil nutrients, and other organisms. Outputs are food and other needs for sustenance with little excess. Conversion of inputs to outputs can never be perfectly efficient so there is inevitable waste, and human consumption itself produces waste. At modest levels, waste is absorbed into the environment, forming a recycling system, more or less at equilibrium. This diagram could apply to any species, including humans as hunters and gatherers.

This steady-state sustainability model is only a first approximation. Paleolithic people did alter and sometimes ruined their natural life support. A prime example is the extinction of the mammoth, a superb resource exploited by ice age humans for its skin, hair, meat, tusks, and bones. The extermination of this species and

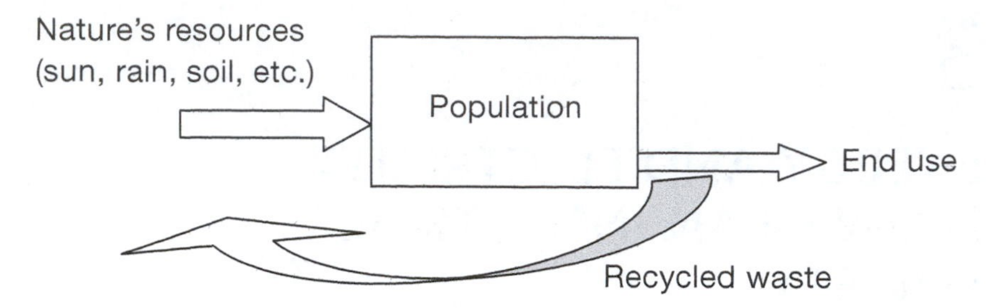

FIGURE 3.1 The carrying capacity model

other megafauna seems at least partly due to human hunters (Martin 2005). Still, for most purposes it is acceptable to assume that small hunter-gatherer populations were sustained but limited by a natural carrying capacity. Some agrarian societies, operating almost entirely on solar-based indigenous renewable resources, may be approximately represented by Figure 3.1, despite instances when agrarian people themselves, or changing climate, depleted their life sustaining environments (Diamond 2005).

In his bestselling book *The Population Bomb* (1968), ecologist Paul Ehrlich, mentioned in the last chapter for his bet with Julian Simon, implicitly assumed the carrying capacity model in raising his alarm about seemingly exponential growth in the human population. He famously but wrongly predicted that millions of people would starve to death in the 1970s and 1980s. This produced spirited debate between neo-Malthusians, led by Ehrlich, and critics more economically and technologically oriented, who pointed out that birth rates were falling in most industrial countries and the largest Third World nations, and that industrial agriculture (the energy-intensive Green Revolution) was raising food yields faster than the population was growing. Later acknowledging that population was not always the central factor of importance, Ehrlich and Holdren (1971) formulated the equation

Environmental Impact = Population*Affluence*Technology

now known as the IPAT model. This was conceptually helpful but difficult to estimate, though Dietz and Rosa (1997) made useful attempts, for example, operationalizing environmental impact as CO_2 emissions and affluence as GDP.

Figure 3.2 illustrates the energy flow through an industrial society, incorporating important new elements. As before, population size is represented by the area of the box with added consideration of the social organization of the population. Affluence may be conceptualized as the size and quality of its inputs (or, alternatively, as its outputs), no longer limited to sunshine and local nutrients but now supplemented (or overwhelmed) by fossil fuels and nuclear power, investment capital, advanced technology, and other industrial resources. This combination removes the cap on natural carrying capacity, so population can be much larger.

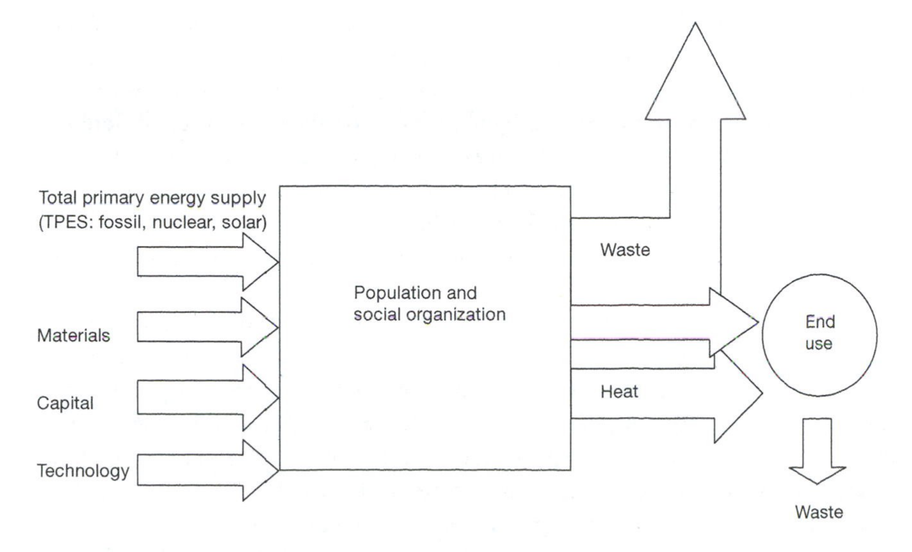

FIGURE 3.2 The industrial society model

Fossil fuels and useful minerals are of course finite, but they are distributed irregularly in the earth, with some deposits easy and inexpensive to reach, while others are difficult or virtually inaccessible until sophisticated new technology and costly investment deliver them for consumption. It is implausible that in the short or medium run humans will literally use up any inanimate resource (as we have mammoths) because there will be remnants that are too difficult or expensive to exploit. The essential problem of high resource consumption is not that we will run out, but that it becomes very costly in money, environmental degradation, accidents, habitat loss, human health, climate change, and wars fought over resources. As long as these costs are borne, there is no easily calculable limit on carrying capacity (Cohen 1996; Kerr 2009).

Nor is industrial society a recycling, steady state system because the inevitable waste occurring at each energy transformation mostly accumulates, often with undesirable effects. Chemical energy in coal is transformed to electrical energy by first burning the coal to heat water until it becomes steam at high pressure, which turns a turbine connected to a generator that produces a voltage difference across a wire. (The generator itself is essentially nothing more than a coil of wire spinning in a magnetic field.) The efficiency of this transformation is typically one-third, i.e., the amount of chemical energy in the coal, released by burning, is three times the amount of electrical energy that comes out of the generator. Energy wasted in this and other transformations is released as heat to the environment, accompanied by polluting residues.

The Second Law of Thermodynamics posits the impossibility of wholly eliminating waste when work is performed in any closed system. As a practical matter, industrial society's waste far exceeds that required by the Second Law, and engineering improvements in efficiency are attainable in all technological applications, though usually at increased cost. Apart from pure technology, existing societal and economic arrangements facilitate if not actually encourage waste, e.g., using a heavy car to transport a far lighter person, or excessive air conditioning in summertime. A new perspective called "ecological modernization," the work of mostly European sociologists, optimistically promotes the reform of market economies toward environmental sustainability while maintaining or enhancing beneficial lifestyles, however there have not yet been great strides in this direction (Mol et al. 2009).

The "black box" model in Figure 3.2 leaves unspecified those processes within the box, though surely the organizations of a large population are crucial to its energy production and consumption. These are best understood as parallel to, or increasingly separate from, the physical flows of energy through the society. So as a prerequisite, we turn to explicit energy paths in a contemporary industrial society, from sources to end uses, as pictured in Figure 3.3. The top row shows the major sources of energy: the fossil fuels (natural gas, coal, oil), nuclear power, and renewable sources (hydro, solar, wind, biomass – all derived from the sun's radiation). To illustrate magnitudes, the total primary energy supply (TPES) for the US in 2010 was 22 percent from natural gas, 22 percent from coal, 45 percent from oil, eight percent from nuclear power, and seven percent from renewables (mostly hydro). Forty percent of TPES was converted to electricity before final use, while most of the remaining 60 percent was used directly as heat, including the explosive heat that drives internal combustion engines.

There are inefficiencies in the various processes that convert primary energy inputs to useful energy outputs (or "end uses"). Recall that the transformation of burnable fuel to electricity is only about one-third efficient. Internal combustion engines are worse, using roughly five percent of the chemical energy in gasoline to actually move the vehicle. Because of inefficiencies – some of them inevitable, some not – only 43 percent of the amount of TPES was directly available to American end users in 2006, about 30 percent of that in the form of electricity. Inefficiencies in the end uses themselves cause further losses.

Figure 3.3 gives a nice qualitative picture of important energy flows. It shows that some natural gas is burned for direct end use (e.g., for space or water heating and for stoves) and some is first burned in power plants, producing electricity that is fed into the power grid and distributed to consumers for other end uses such as lighting and air conditioning. At one time the situation for coal and oil was similar. Today, virtually all coal is converted to electricity at large generating stations, where air pollutants (except carbon dioxide) are fairly well controlled. The US gets nearly half its electricity from coal and about 15 percent from natural gas. Nuclear power is used only for electricity generation, supplying 20 percent of US electricity. Hydroelectricity is presently the only renewable making a substantial contribution to the American energy supply.

Oil is a complex mixture of hydrocarbons. Crude petroleum is sent to a refinery that separates the various fractions including gasoline, diesel fuel, and jet fuel. Today

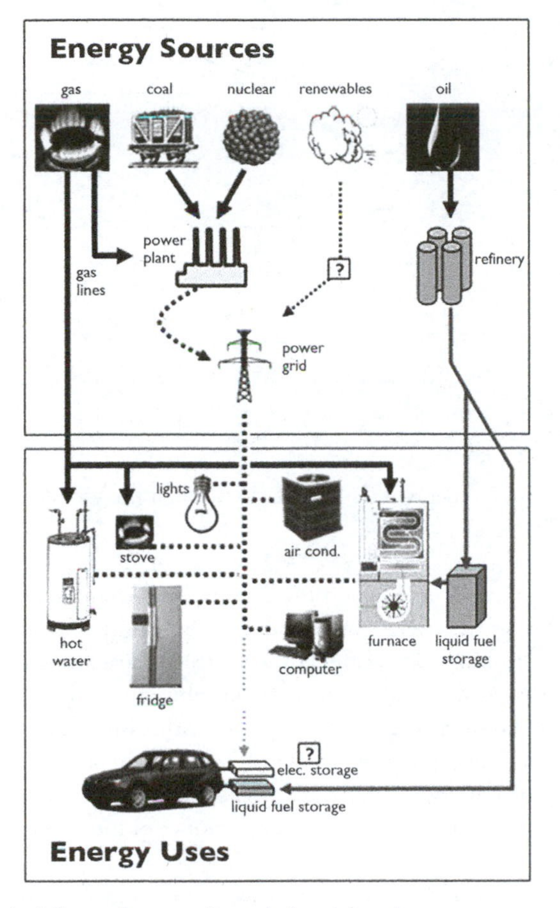

FIGURE 3.3 Physical flow of energy in an industrial society

almost all oil is used for transportation, and most transportation depends on oil. (Petroleum is also an important feedstock for the chemical industry, but here we are concerned only with its use as a fuel.)

Figure 3.4 shows more quantitatively but abstractly the *primary energy* flow through the United States in 2010. Produced by the US Energy Information Agency, energy quantities are here expressed in quadrillion BTUs (Quads). Unfortunately the diagram does not show the conversion of primary fuels into electricity, which considerably lessens the amount of energy that actually reaches end users. (The typical electrical power plant, operating at one-third efficiency, requires three units of primary fuel to produce one unit of electrical output.)

Trading energy

Hughes (1983) emphasized the importance of corporations, government agencies, and the people who run these organizations – rather than the physical character

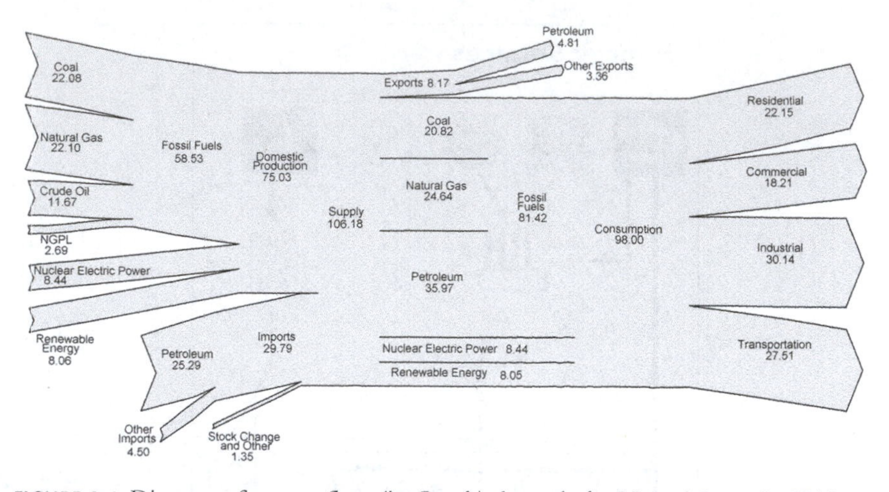

FIGURE 3.4 Diagram of energy flow (in Quads) through the United States in 2010 (EIA)

of a fuel or electricity – in shaping the energy industries and their momentum toward increased growth and consumption. The business strategy of John D. Rockefeller in building the Standard Oil trust, apart from eliminating all competition, was to vertically integrate, combining in one corporation the exploration, extraction and refining functions with distribution of products and ownership of retail outlets. Following Rockefeller, the major energy corporations moved entirely or considerably toward vertical integration with the caveat that government prevented any one from holding monopolistic control over an energy sector.

When components of the supply chain were not connected by unified (private or government) ownership, their connections could be solidified by long-term contracts. An electric utility company might ensure guaranteed supply and price of fuel (say, coal or natural gas) and its delivery to the generating station (via railroad or pipeline) through 20-year contracts, or might simply purchase a coal mine and railroad. Sales of fuels or electricity to very large end users would also be assured by long-term contracts. This consolidation of links in the supply chain allowed economies of scale, quantity discounts, eased planning for future expansion, and ensured the reliability of energy supply.

For most of the twentieth century, the social arrangements and institutions of energy – including labor-management relations, government regulation of (or subsidies to) industry, sales promotion, lobbying, etc. – largely mapped onto the physical supply chains shown in Figure 3.3, or to a particular location in the chain. In other words, relevant organizations and social relationships more or less corresponded to the physical flow of energy in each sector.

That changed after the 1970s. Increasingly, the trading of fuels and electricity as fungible products placed control of the energy system in the hands of people and organizations that never handle or directly deal with the physical flows that are diagramed in Figure 3.3. Sometimes referred to as "restructuring," this change began after several shocks independently shook up what seemed, at the time, an inexorably growing complex of energy industries. OPEC, exploiting the Yom Kippur War of 1973 and the Iranian Revolution of 1979, finally gained enough muscle to constrain the supply of oil on the world market and force up the price of its exports (though the price of oil collapsed in the early 1980s before rising again). Some oil exporting countries nationalized the concessions held by Western oil companies. At the same time, the heretofore complacent growth of the US electric power industry suffered multiple setbacks including huge blackouts (in 1965 and 2003); an overly optimistic commitment to very large generating stations that did not perform as well as expected; and public opposition to – and stricter regulation of – nuclear power, which, combined with high inflation of the 1970s, drove the customarily decreasing price of electricity sharply upward.

Also at that time was a strengthening reaction among conservative politicians and economists against government control and regulation of industry, urging its replacement by privatization and free markets. This took hold first in Britain during the administration of Margaret Thatcher, followed by the United States during the presidency of Ronald Reagan. Regulated business activities including long-distance telephone calls, airline ticketing, and financial services were deregulated, initially with the benefits of lower prices and improved service (Yergin & Stanislaw 2002). With these successes, the energy industries looked ripe for deregulation.

In the 1970s there was hardly a world market in which barrels of oil were traded back and forth. The transfers that did take place were mostly within vertically integrated oil companies, moving oil from the well, into tankers, to refineries, to gas stations. By the 1980s there was a world market for oil, its price determined through bargaining between buyers and sellers, and the most important marketplace the New York Mercantile Exchange (NYMEX), where agricultural products had been traded since the nineteenth century. "Now the price of oil was being set by the interaction of the floor traders at the NYMEX with other traders and hedgers and speculators all over the world. Thus was the beginning of [trading in] the 'paper barrel' "(Yergin 2011: 167).

For an airline or independent oil producer to purchase "futures" – guarantees of price at some future date to protect itself against a rise in price – someone had to be on the other side of the trade. That was the speculator, who never took delivery of the oil but only wanted a profit on the trade by successfully anticipating rises and falls in price.

> Oil was no longer just a physical commodity, required to fuel cars and airplanes. It really had become something new – and much more abstract. Now these paper barrels were also, in the form of futures and derivatives, a financial instrument, a financial asset . . . Economic growth and financialization soon came together to start lifting the oil price higher. With that came more volatility, more fluctuations in the price, which was drawing in the traders. These were the nimble players who would, with hair-trigger

timing, dart in and out to take advantage of the smallest anomalies and mispricings within these markets.

(Yergin 2011: 171)

It became difficult to tell the extent to which spikes or drops in the price of a barrel of oil might be caused by, say, changing demand in China or the manipulation of speculators. Depending on market conditions, the cost of gasoline might be only loosely tied to the cost of crude oil (Lefebvre 2011).

In the US during the 1970s the federal government held the price of natural gas low despite constriction of supply. Here was another commodity ripe for deregulation. Freeing the price to rise with demand motivated gas companies to bringing larger supplies to market, and after deregulation, natural gas became plentiful. The Enron Corporation, a major innovator in natural gas marketing, is today known as an emblem of financial collapse and fraud, its 2001 bankruptcy then the largest in US history, its chief executives sent to jail. But a decade earlier Enron was among the nation's highest flying corporations, named by *Fortune* magazine as "America's Most Innovative Company" for six consecutive years, its stock price continually rising. The crux of Enron's early success was its application of oil's trading model to natural gas, buying and selling "paper therms." The tools of the trader came into play: futures contracts, short selling, derivatives, arbitrage, credit default swaps, and so on.

Later Enron became one of the first and largest electricity traders, a more complex business, requiring a dismantling of the "natural monopolies" that had been the basis of the private utility industry in the US. Breaking down the long-standing vertical integration of regional utility companies required that ownership of electrical transmission and distribution lines be separate from ownership of electrical generators, so that different generating companies, competing for the same customers, all had access to the region's transmission and delivery network. The extent and form of deregulation varied from state to state. California's electricity crisis of 2000–01 was the result of especially inept implementation that deregulated the price of wholesale electricity but not of retail sales, so that the state's utilities were forced to buy electricity from independent generators and traders (including Enron) at far higher prices than they could charge end users, producing bankruptcy or near-bankruptcy of the state's utilities, electricity shortages, and rolling blackouts. These were not caused by any physical lack of fuels or electricity, but by muddled (and sometimes fraudulent) profit-making in the deregulated trading system. This experience put a cautionary hold on deregulation, later reinforced by the disastrous collapse of the US financial system in 2008, partly the result of deregulation.

It is uncertain if there will be further deregulation of electricity, but certainly trading is here to stay as a mediator between the input of primary fuels and the output to end users. Yet even some sophisticated analyses of global energy virtually ignore the marketers that sit atop the flows of oil, natural gas, and electricity. Perhaps this is because the traders do not easily map onto the physical flow of fuels and electricity, having no more "hands on" contact with their commodity than the agricultural products traders, longer residents of the NYMEX, had with butter and cheese.

In some ways, speculation on energy futures is a good thing, allowing energy producers, wholesalers, and consumers to "hedge" their positions more efficiently, protecting themselves against unforeseen shifts in energy prices. But it is one thing to have legitimate participants in the energy business place strategic bets on where prices will be months into the future. It's another thing to have hedge funds and banks that do not really participate in the energy industries pump billions of purely speculative dollars into commodity exchanges, driving up gasoline or electricity prices, their only purpose being to increase their trading profits at the expense of everyone else. That distinction is why the US government placed limits on pure speculation in grain exchanges after repeated manipulations of crop prices during the Great Depression.

In 1991, a few years after oil futures began trading on the NYMEX, the US Commodity Futures Trading Commission allowed Goldman Sachs to process billions of dollars in speculative oil trades. By 2008, eight investment banks accounted for 32 percent of the total oil futures market. Only about 30 percent of oil futures traders are actually oil industry participants. Those eight banks alone can severely inflate the price of oil (Kennedy 2012).

Trends in energy and electricity consumption in industrial nations

The Organization for Economic Cooperation and Development (OECD) comprises nearly a fifth of the world population, its roster roughly corresponding to a list of industrial nations. The OECD consumes over two and a half times as much energy per capita as the world average. This is due partly to extensive electrification, a high rate of cars per household, large industry and service sectors, high heating degree-days, and high GDP per capita. On the other hand, OECD nations use only half the energy required by the rest of the world to produce one unit of GDP, a consequence of technological efficiency in the energy transformation sector (especially power plants) and in final consumption (better mileage cars, insulation of buildings), plus the movement of high-energy-consuming industries to the Third World (IEA 2009).

Following United Nations statistical usage, I count the industrialized or "developed" nations as Japan in Asia; Canada and the United States in North America; Australia and New Zealand in Oceania; and Europe, excluding the former communist nations of Eastern Europe. I limit this initial analysis to 21 OECD nations larger than two million in current population because small countries (Iceland and Luxembourg) add little to the overall picture. Detailed analysis focuses on energy trends in eight of the largest industrial countries: the United States, Canada, United Kingdom, France, Spain, Italy, Japan, and Australia. These are diverse in size, location, climate, language, and culture, hence providing a good overview of industrial democracy. A conspicuous omission is Germany, where reunification in

1990 was so great a structural change that comparison of its energy trends to the other cases makes little sense.

Energy data used here are from "Energy Balances of OECD Countries (2009 Edition)," available by subscription to the International Energy Agency (IEA) (http://wds.iea.org/wds/ReportFolders/ReportFolders.aspx), which provided the longest time series (since 1960) of energy trends that is reasonably comparable across nations.

The industrial nations vary considerably in per capita TPES and electricity. The highest users are in North America and Scandinavia, the lowest are near the Mediterranean (Table 3.1). All industrial nations are high consumers by Third World standards. Per capita, even Portugal uses twice the energy and electricity of China, and seven to ten times as much as India.

The Mediterranean climate requires less heating than northern Europe or most of North America. Normalization on climate reduces the disparity, but differences remain large (Smil 2009). The United States, Japan, urban Australia, and New Zealand are in temperate zones and on average (recognizing local variation) do

TABLE 3.1 TPES per capita (most to least), electricity consumption per capita, and population in 2008 (IEA)

Nation	*Population (millions)*	*TPES per capita (gj)*	*Electricity consumption per capita (gj)*
Canada	33	338	59
USA	304	315	49
Norway	5	277	91
Finland	5	276	59
Australia	21	256	42
Belgium	11	228	31
Sweden	9	226	53
Netherlands	16	202	26
France	62	175	28
New Zealand	4	171	35
Germany	82	170	26
Austria	8	164	30
Japan	128	161	29
Switzerland	8	149	31
Denmark	6	147	24
Ireland	4	143	22
UK	61	142	22
Spain	46	126	22
Italy	59	124	21
Greece	11	122	19
Portugal	11	97	18
Median:	**11**	**170**	**29**
Mean:	**43**	**191**	**35**

not require inordinate heating or cooling. Most of Canada's population resides in southern Ontario and Quebec, suffering colder temperatures than most of the United States, which may account for higher per capita energy and electricity consumption in Canada than the US. Lower energy use among the southern European nations is also attributable to their late industrialization.

I will focus on the eight countries with current populations over 20 million (excluding reunified Germany), which together account for over 80 percent of the TPES and electricity used by the developed world. Trends in per capita consumption of TPES and of electricity, from 1960 to 2008, are shown in Figures 3.5 and 3.6, respectively. Canada and the United States are by far the highest per capita users of primary energy, and Canada outstrips even the US in per capita use of electricity. These two nations have similar cultures and an interconnected energy system, so the similarity of their trends is expected. Australia occupies a middle position. The remaining nations – the Europeans and Japan – are clustered at the lower end of the range (with Spain and Italy lowest).

The rank order of TPES or electricity consumption in 2008 is fairly predictable from the rank orders in 1960. Australia is an exception, a higher consumer in 2008 than one might have expected from its place in 1960. The United Kingdom is another, now a lower user than would have been guessed in 1960. Still, the general pattern is clear: nations that consumed the most (or least) energy and electricity in 1960 continued to do so in later years. Each nation had a momentum – whether

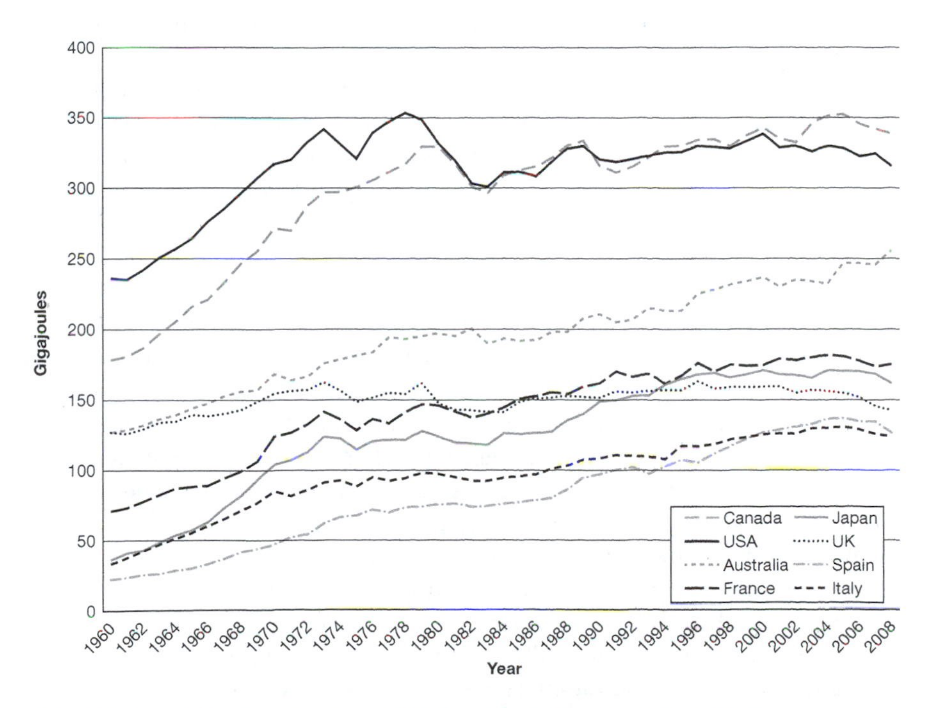

FIGURE 3.5 TPES per capita since 1960 for eight industrial nations (IEA)

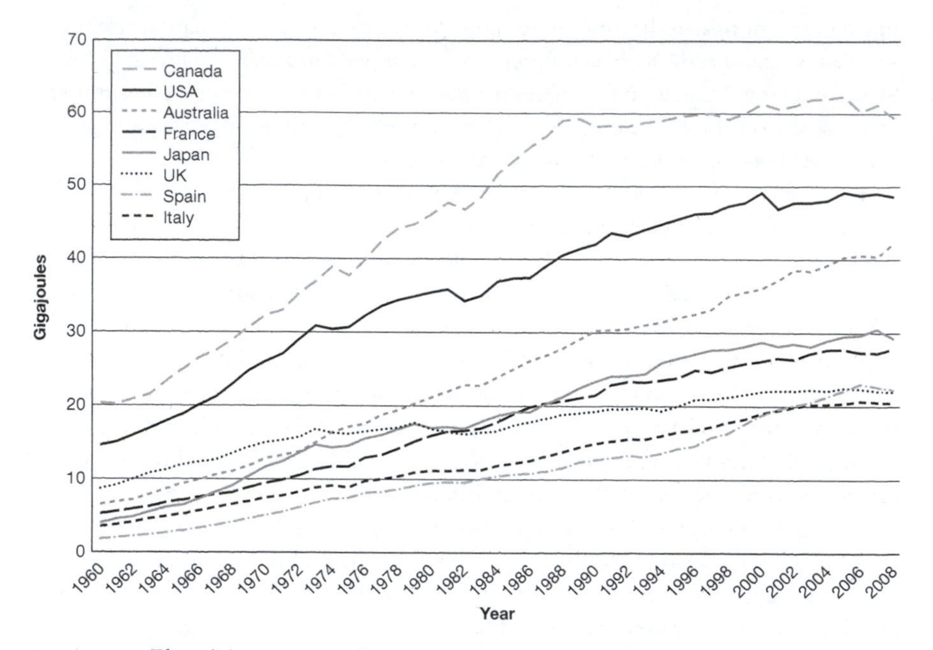

FIGURE 3.6 Electricity consumption per capita since 1960 for eight industrial nations (IEA)

literal or figurative – that carried it through the decades since 1960 on a persistently upward course.

The 1973 OPEC oil embargo and the 1979 Iranian revolution were followed by sharp rises in oil (and other fuel) prices (Figure 2.2). Both events left salient marks in the Canadian and American trends and to a lesser extent in the graphs of most other nations.

Conclusion

Patterns of energy and electricity consumption beg for sociological analysis. I show as a prerequisite the complex flows of energy through an industrial society, avoiding naïve missteps such as the notion of a fixed carrying capacity, or ignoring the energy organizations that bring fuels from the ground to end users. Industrial societies are flexible, capable of mobilizing technology to use diverse resources and, when necessary, to substitute one resource of end use for another, but always at a cost and by producing waste. Our essential problem is that we pay excessive social, environmental, and monetary costs for excessive use of fuels.

In industrial nations, with a few exceptions, per capita energy consumption (TPES) increased faster than population growth, at least since 1960. The oil shocks interrupted these trends, especially in petroleum consumption, though often consumption turned upward again. However in the US and Canada, the oil shocks instigated a longer period of relative leveling in TPES per capita.

The situation of electricity is different, rising with faster momentum and showing little halt with the oil shocks of the 1970s. Per capita electricity consumption rose, on average, twice as fast as general energy consumption, and nearly five times faster than population growth. Increasingly, the industrial nations are using their primary fuels to generate electricity. Given current trends, the United States will soon devote the majority of its fuels to the production of electricity. Since two-thirds of the energy in fuels is lost as waste heat during the process of electricity generation, this may be an inefficient use of resources.

Supplementing these trends with other information, here is a narrative for the period since World War II: In the first postwar decades there were very high (and wasteful) increases in energy consumption and electrification, both in countries that had been destroyed and were rebuilding, and in the undamaged nations of North America, enjoying a postwar boom. This period ended in 1973 when OPEC embargoed petroleum to nations supporting Israel in the Yom Kippur War, producing sharp rises in energy prices and long lines at gas stations, the so-called "energy crisis." The Iranian Revolution of 1979 under the Ayatollah Khomeini sent energy prices far higher. (The shocks of 1973 and 1979 reduced energy consumption in Britain more than in any other nation, while at the other extreme, Australia barely deviated from its upward path.) High energy prices motivated conservation and the development of new energy supplies. These curatives turned the world's oil scarcity into a glut, sending prices downward, undermining OPEC (and also the Soviet Union, which depended heavily on revenue from oil exports). With lower prices in the 1980s, demand increased in the industrial nations and also importantly in China and India, turning prices upward again. Prices continue to fluctuate, probably reflecting speculative trading on global energy markets.

The rank ordering of nations in per capita consumption of energy and electricity was set by 1960 (probably earlier) and persists. Once on its energy trajectory as a high or low consumer, with corresponding energy prices (gasoline including taxes is much cheaper in North America and Australia than in Japan and Europe), each nation continued along that path. This finding is consistent with historian Thomas Hughes's (1983) hypothesis that the energy industries of each nation have their own momentum, ensuring continued growth. Possibly this "momentum" is no more than a figurative construct imposed upon trend lines, but more likely it is a real societal mechanism, as Hughes meant it. In each nation, each fuel and the electric grid have a nexus of energy producing and consuming organizations, corporate and governmental, that keeps its energy flowing and fosters growing consumption, with an explicit goal of sustaining and enriching the organizations themselves.

4

ENERGY, ELECTRICITY, AND QUALITY OF LIFE

The remarkable improvements in quality of life that occurred during the industrialization of Europe, North America, and Japan in the nineteenth and early twentieth centuries were caused, in large part, by the invention and adoption of energy-intensive technologies (Smil 2005). Coal was used to fuel steam engines, petroleum fed internal combustion engines, and all the fossil fuels plus falling water turned electrical generators.

It is important to distinguish electricity consumption from total energy consumption because rapid and pervasive electrification has been the hallmark of industrialization (Hughes 1983; Platt 1991; Nye 1992). Energy in the form of electricity is versatile, essential for modern technologies, and regarded as of higher quality than directly burned fossil fuel. Since commercial electricity's beginning, people have seen the mysterious force as intrinsically modern and desirable. Lenin (1920) famously declared that Communism was Soviet power plus the electrification of the entire country. In Depression-era America, the electrification (and flood control) of the Tennessee Valley epitomized progressive governmental social policy.

Industrialized nations surpass the Third World on many desirable features of life including lower infant mortality, gender equality, ample food and clean water, civil rights, and the stabilization of population size. It has long been noted that per capita energy and electricity consumption are highly correlated with economic development and other indicators of modern lifestyle, with the inference that the more energy that is consumed, especially in the form of electricity, the better life is (e.g., Starr 1972; Erol & Yu 1987).

Despite continuing suggestions that increasing consumption of energy and electricity will improve the wellbeing of industrial societies, there are reasons to doubt this will occur in already highly consuming nations. It is profitable for industries that supply energy to encourage high consumption, whether essential or not. The classic American example of wasteful promotion is the "all electric house," which evoked a sense of modernity but was inefficient when using natural gas to generate electricity for cooking and space heating, rather than burning gas directly in stoves and furnaces.

Another reason that increased energy consumption may not bring much benefit is that today's industrial societies enjoy levels of wellbeing that in some respects approach saturation. Nearly all citizens of developed societies have easy access to clean water, adequate housing, dietary calories, and electrical appliances. Infant mortality rate is nearly as low as it can go. For these nearly "maxed out" aspects of wellbeing there is little room for improvement, no matter how much additional energy is consumed.

During and after the 1970s, as energy problems came to the fore in both domestic and foreign policy, the presumed causal relationship between energy/electricity use and societal wellbeing was studied more closely and tentatively qualified. There is no doubt that increased energy consumption is essential for the development of poor societies. But once nations reach the high plateau of energy and electricity consumption that is characteristic of developed economies, additional usage seemed to be only modestly if at all associated with further improvement in wellbeing (Mazur & Rosa 1974; Schipper & Lichtenberg 1976; Darmstadter et al. 1977; Rosa et al. 1980; Smil 2003; Dietz et al. 2008). These studies were based on cross-sectional analyses, a weak foundation for causal inference.

With the dramatic changes of recent decades, this core issue requires further examination. For the first time we can address this question using longitudinal data: Did changes in energy or electricity consumption since 1980 affect material wellbeing in industrialized nations?

Methodological issues

As in the prior chapter, this analysis is based on 21 industrialized nations of the OECD with populations of over two million. Here energy statistics by nation and year from 1980 onward are taken from the US Energy Information Agency (http://www.eia.doe.gov/iea/wec.html; EIA, 2008). Net electricity consumption, which excludes energy consumed by the generating units, is total net electricity generation + electricity imports – electricity exports – electricity distribution losses. It is measured in megawatt-hours (MWh) and hereafter simply called electricity consumption. Total primary energy consumption includes all of the energy types (fossil fuels, nuclear, hydro and other renewals and geothermal) + electricity imports – electricity exports.

There is no fully satisfactory way to measure the wellbeing of a nation's people, and considerable attention has been devoted to this problem (McGillivray 2006). In 1990 the United Nations adopted the Human Development Index (HDI), combining three indicators: life expectancy, adult literacy rate (and school enrollment), and GDP per capita. The HDI has the advantage of a simple summary variable, but since its components are not measured in commensurable units, they are necessarily combined in an arbitrary way.

Following earlier multiple-indicator studies and constrained by the availability of comparative data, I selected 13 variables that measure diverse aspects of wellbeing and are related to per capita consumption of energy and electricity among the world's nations. These are life expectancy and (somewhat redundantly) infant mortality rate; physicians and hospital beds per capita; rate of enrollment in college (tertiary education); internet users per capita; fixed and mobile phone subscribers per capita; percent of households with television; passenger cars per capita; GDP per capita (based on purchasing power parity, in constant international dollars); male suicides per capita; divorce rate; and percentage of population satisfied with their lives.

Every indicator is assailable as a normative measure of well- being. High GDP per capita or a high number of cars (or other consumer goods) per capita is especially subject to the criticism of wastefulness. Longer life expectancy has diminishing returns when it is accompanied by loss of vigor, disability and depression, requiring social services for elderly citizens who are no longer productive and who no longer enjoy their lives. Divorce, usually considered the failure of a marriage, may be regarded as freedom to end an unhappy relationship. A low suicide rate might reflect the forbiddance of suffering people at the end of life to induce their own "death with dignity." Even decreasing infant mortality, to which hardly anyone would object, raises population size and hence environmental load. There is also the matter of how equitably benefits are distributed through a society, which is not captured by these variables. With such qualifications, it is still true that in the industrial democracies, the focus of this inquiry, there is general agreement that people are better off when each indicator moves one way rather than the other. If the same finding reliably occurs across a broad range of indicators, then misgivings one might have for any particular indicator become less weighty.

Most indicators used here are listed in the World Bank's *World Development Index Online* (2009) for as recently as 2006. For developed nations, most trends go back to 1980 or earlier. Occasionally missing values for a year are interpolated from adjacent years. Suicides per capita are in the *United Nations Demographic Yearbook* (1985, 2006). Until the unification of Germany in 1990, data for East and West Germany are combined.

College enrollment is represented by the World Bank variable, school enrollment, tertiary (gross percentage), defined as the number of students enrolled in tertiary education (regardless of their age), as a percentage of the age group that officially corresponds to tertiary education (in the US, ages 18–22). Graduate students and undergraduates older than 22 are counted in the numerator but not the denominator, so this is an inflated measure of enrollment.

A subjective measure of wellbeing is taken from the World Values Survey, which compiles surveys from 81 nations. Following Inglehart et al. (2008), I use the question, "All things considered, how satisfied are you with your life as a whole these days?" Respondents answered on a 10-point scale from 1 = dissatisfied to

10 = satisfied. This question is available for 15 of my nations from the period 2005 to 2007, and for 13 of these during the period 1981 to 1990, so change is measurable. I counted the percentage responding 9 or 10 as satisfied.

To date, the hypothesis that increased energy (or electricity) consumption produces improved quality of life has been tested only with cross-sectional data. Knowing changes in 11 of the indicators of quality of life, from approximately 1980 (1990 for internet usage) to 2006, allows the first test with longitudinal data.

My analytical approach is based on correlations, which do not provide direct evidence for causal relations. However, we may reasonably infer from the lack of a correlation between two variables that probably one is *not* the cause of the other. Uncertainty in reporting yearly energy consumption makes it difficult to distinguish signal from noise over short time intervals, so I focus on change over a quarter century. I tested shorter timespans with similar results.

Looking at all the world's nations, distributions of variables are often highly skewed with extreme outliers. Limiting the analysis to industrial democracies, which in many ways are similar, eliminates most outliers and skew. Therefore, correlations are calculated on untransformed data. Correlations after log transformations are essentially unchanged.

The propriety of using significance tests is dubious since the 21 nations do not comprise a random sample; also sample sizes vary depending on the indicator. Nonetheless, significance levels are often requested so I report them with the caveat that p values serve more as benchmarks than as true probabilities.

Results

Energy use is strongly correlated with diverse indicators of quality of life among the world's nations. Figure 4.1 is an illustration, showing life expectancy in 2006 as a function of per capita energy consumption for 135 countries with populations over two million. Nearly all nations using above 40 megawatt-hours per person have life expectancies near 80 years, and consumption twice as high produces no increase in longevity. The scattergram is fit with a logarithmic curve ($R^2 = 0.69$), showing the leveling of longevity at higher levels of consumption. The left portion of the figure shows life expectancy as a function of per capita electricity consumption, again producing a logarithmic relationship ($R^2 = 0.70$). Nations using more than five megawatt-hours of electricity per capita have life expectancies around 80 years; there is no further improvement with up to three times more electrical usage. Even Portugal, the lowest per capita consumer of energy and electricity, has reached the inflection points.

Table 4.1 gives means, medians and ranges for 13 life quality indicators in the industrialized nations in 2006. For each indicator, the median and mean are fairly close, suggesting no major skews, which I verified by inspection.

Correlations between energy or electrical consumption (per capita) and these life quality indicators are shown in Table 4.2. GDP per capita and rate of internet usage both increase with energy and electricity per capita. Percentages enrolled in

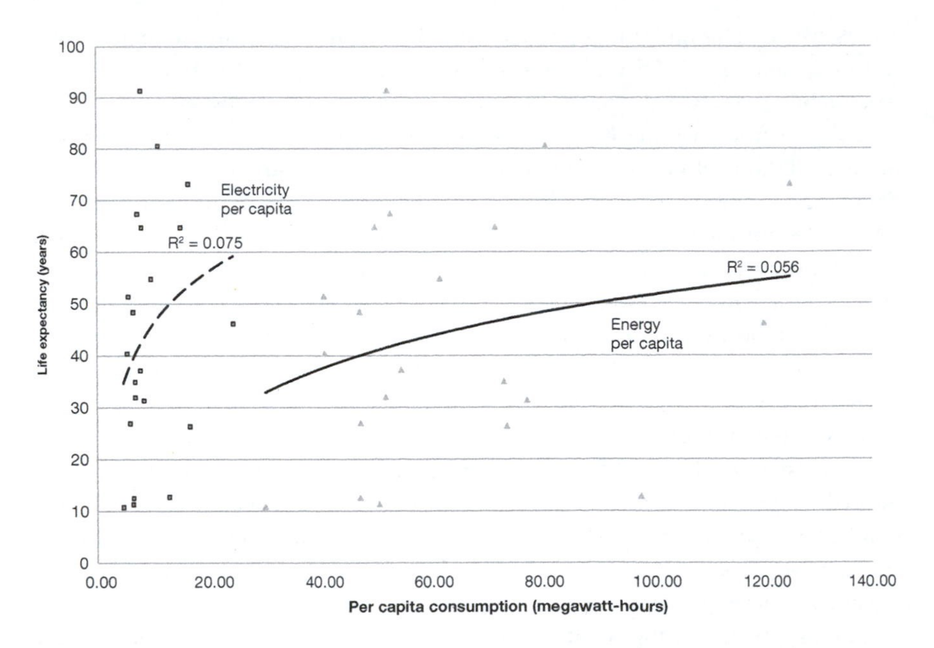

FIGURE 4.1 Life expectancy increases with per capita energy consumption for 135 countries (EIA, World Bank)

TABLE 4.1 Means, medians, and ranges for 13 quality of life indicators in the industrialized nations, 2006 (n = 17–21)

Indicator	*Mean*	*Median*	*Range*
Life expectancy (years)	79.9	79.8	77.8–82.3
Infant mortality (per 103 births)	4.0	3.7	2.6–6.5
Physicians (per 103 popn.)	3.2	3.4	1.9–5.0
Hospital beds (per 103 popn.)	5.6	5.0	3.3–14.7
Male suicide rate (per 105 popn.)	18.6	18.1	6.3–35.5
Divorce rate (per 105 popn.)	2.3	2.4	0.8–3.6
Percent in college	68.3	67.0	45.8–94.9
Percent households with TV	95.6	97.6	79.1–99.0
Fixed/cell phones (per 102 popn.)	151	152	122–179
Passenger cars (per 103 popn.)	481	461	354–607
Internet users (per 102 popn.)	60.9	62.6	18.4–85.6
GDP per capita (ppp)	$32,999	$32,071	$20,151–48,526
Percent satisfied with life	27	26	15–42

TABLE 4.2 Correlations between energy or electricity consumption (per capita) and 13 indicators of quality of life

Indicator	*Consumption of primary energy per capita*	*Consumption of electricity per capita*
Life expectancy	0.00	0.14
Infant mortality	–0.30	0.00
Physicians per capita	–0.48	–0.33
Hospital beds per capita	–0.14	–0.14
Male suicide rate	0.17	0.17
Divorce rate	0.47	0.31
Percent in college	0.28	**0.37**
Percent households with TV	0.10	0.00
Phones per capita	–0.52	–0.33
Passenger cars per capita	0.14	0.00
Internet users per capita	**0.60**	**0.59**
GDP per capita	**0.59**	**0.62**
Percent satisfied with life	0.33	**0.55**

Bold indicates significant improvement with higher consumption ($p < .05$, one tail).

college and satisfied with life increase only with electricity consumption. Most correlations in Table 4.2 are small and insignificant. Some moderate correlations go the "wrong" way: high energy or electrical consumption per capita is associated with fewer physicians per capita, fewer telephones per capita, and higher divorce rate.

Cross-sectional data permit only a very weak test of the hypothesis that higher energy consumption causes improved wellbeing. It is preferable to compare change in energy consumption over time with corresponding changes in the quality of life. There are numerous determinants of energy/electricity consumption from nation to nation, and year to year, including relative cost of energy, patterns of imports and exports, and changing structure of an economy, but overall, if increasing consumption of per capita energy/electricity causes improvement in wellbeing, this ought to emerge as a signal over a period as long as a quarter of a century. Again using life expectancy as an example, Figure 4.2 represents each of the 21 developed nations with an arrow running from 1980 to 2006. Each arrow begins at that nation's coordinates of the graph for energy consumption (per capita) and life expectancy in 1980. The arrow ends at the coordinates for 2006.

All arrows point upward, showing that life expectancy improved considerably everywhere from 1980 to 2006; it is not a "saturated" indicator. Changes in per capita energy consumption were highly variable, with five nations slightly decreasing their per capita usage, another remaining nearly the same, and the rest increasing. Overall, life expectancy rose irrespective of whether nations increased or decreased energy consumption per capita. Whatever caused this enhanced longevity, it was not in general an increase in the combustion of fuel.

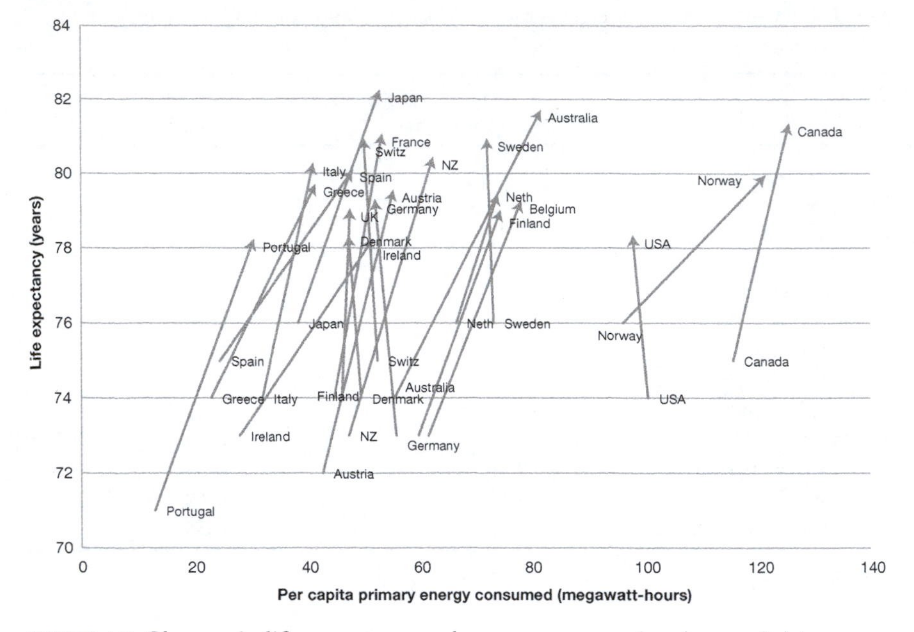

FIGURE 4.2 Changes in life expectancy and energy consumption (per capita) in industrial nations, 1980–2006 (EIA, World Bank)

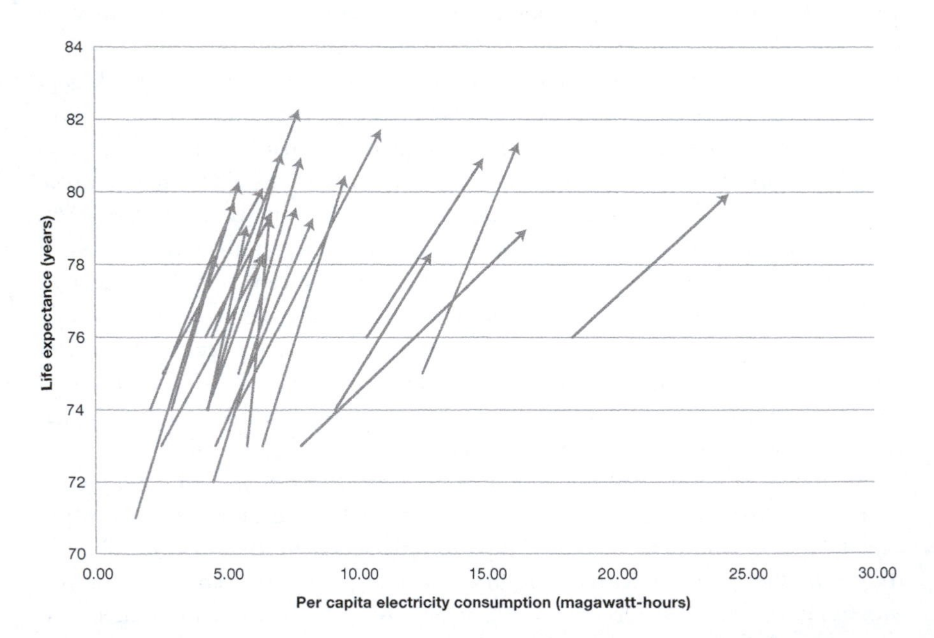

FIGURE 4.3 Changes in life expectancy and electricity consumption (per capita) in industrial nations, 1980–2006 (EIA, World Bank)

TABLE 4.3 Correlations between changes in energy or electricity consumption (per capita) and changes in 11 indicators of quality of life

	Change in consumption of primary energy per capita	*Change in consumption of electricity per capita*
Life expectancy	0.27	0.00
Infant mortality	–0.24	0.22
Physicians per capita	0.20	–0.10
Hospital beds per capita	0.08	–0.53
Divorce rate	0.32	0.00
Percent households with TV	**0.53**	–0.10
Phones per capita	0.00	–0.33
Passenger cars per capita	0.37	–0.06
Internet users per capita	–0.32	0.10
GDP per capita	0.28	0.32
Percent satisfied with life	0.00	0.20

Bold indicates significant improvement with higher consumption ($p < 0.05$, one tail).

Figure 4.3 is a similar display except now showing change in life expectancy as a function of changing electricity consumption from 1980 to 2006. It is noteworthy that electricity consumption per capita rose in every nation, even those that reduced their per capita energy usage. All arrows in Figure 4.3 point to the upper right, the result of across the board increases in both electricity consumption and longevity. However, increase in one variable is not proportional to increase in the other ($r = 0.0$), discrediting the hypothesis that the increased electrification contributed directly to increased longevity.

Adequate data for 1980 (1990 for internet use) are available for 11 wellbeing indicators in most nations. Table 4.3 shows correlations between *changes* in each indicator and *changes* in (per capita) energy and electricity consumption, over this quarter of a century. Nearly all correlations are small and insignificant. One exception is a sizable positive relationship between growth in energy consumption and increase in households having television. The cross-sectional links to energy or electricity consumption in Table 4.2 (for internet use, GDP per capita, college enrollment, and satisfaction with life) are not sustained with longitudinal data.

Conclusion

Serious analysts agree that increased energy and electricity consumption is essential for improving the wellbeing of people in less developed nations. China and India, because of their population sizes and rapid growth, are the most salient examples of agrarian nations rapidly increasing their use of fuels and hydroelectricity to spur their economies, create jobs, and improve the material lives of their populations.

On a per capita basis, the wealthy industrial nations still consume far more energy and electricity than the fastest growing nonindustrial societies (excepting the United Arab Emirates and Kuwait). A minority of industrial nations, including the United States, has slightly reduced per capita energy consumption since the 1970s. But even among these nations, per capita electricity consumption continually increases, as does total energy consumption.

Prior analyses based on cross-sectional data have for years hinted that these rich nations may have reached a plateau upon which further increases in energy and electricity consumption produce little if any benefits for quality of life. The present analysis, the first based on longitudinal data, strongly fortifies that supposition. Among industrial nations, increases in per capita energy and electricity consumption over the quarter century between 1980 and 2006 were not accompanied by corresponding improvements in indicators of quality of life. The once common claim that energy and electricity consumption improves per capita GDP, which recent analysts have severely qualified (Erol & Yu 1987; Ghali & El-Sakka 2004), receives no support from this cross-national comparison of trends.

Why, in rich nations, does increasing energy/electricity consumption fail to cause commensurate improvements in quality of life? Some aspects of wellbeing may have improved so much during the past century that there is little room for further gain. With most houses having a television, this indicator is nearly saturated. Infant mortality rate cannot go much lower than its present average of four deaths per thousand births across the industrial nations. But several years were added to life expectancy between 1980 and 2006, suggesting that it may continue to improve (Figures 4.2 and 4.3). Indeed, most of the indicators of wellbeing used here have no near-term limits, and still they are not related to increased energy or electricity consumption. Perhaps longevity would actually improve from a decrease in pollution that would accompany reduced energy consumption.

Lacking a full understanding of causal connections, we cannot say with certainty that reducing energy consumption would have no negative impact, perhaps through macroeconomic effects. But it does seem defensible to assert that the industrial nations consuming the most energy and electricity per capita have plenty of slack that could be tightened without diminishing the wellbeing of their citizens.

Affluent societies consume energy in extravagant and inefficient ways. Of course, there will be specific new uses of energy that have undeniable value for some users. But often the perceived utility of an innovation is an illusion of consumer culture. This is especially true for electrification, one reason that electricity consumption increases faster than nonelectric consumption.

This analysis takes no account of differences among industrialized nations such as climate and geography that may affect the utility obtained from a unit of energy consumed. The Mediterranean climate requires less heating than northern Europe or most of North America, but Smil (2009) claims that controlling on climate explains little difference in overall energy consumption across Europe. The United States, Japan, urban Australia, and New Zealand are in temperate zones and on average (recognizing local variation) do not require inordinate heating or cooling.

Other possibly relevant factors include a nation's endowment with energy sources and historical pattern of energy use. Each nation is here treated without regard to its internal energy flows or its interdependence with other nations, for example, the intimate interconnection between the United States and Canada of both fuel exports and electrical grids.

Why do we in the industrial nations use so much energy and electricity if it does not directly improve our quality of life? Obviously consumer demand is one cause, especially when the inflation-adjusted prices of fuels and electricity remain tolerable or even decrease, perhaps because of government subsidies or increasing efficiency in extraction and utilization.

The most sociologically potent theoretical explanation of continually increasing energy use came from historian Thomas Hughes (1983) in his study of early electrification in the United States, Britain, and Germany. In each nation, the people and institutions involved with electricity, whatever its industrial and institutional form, consistently pursued growth, comprising a kind of social "momentum" arising from the involvement of people whose professional skills are committed to the system. Corporations, government agencies, professional associations, and universities that profited from electrification, or were otherwise invested in it, provided – and still provide – the momentum. From this perspective, growing consumption is caused less by the pull of consumers than the push of institutions and industries that produce the fuels and electricity, or sell the products that use them.

PART II

Energy sources and consumption: using more, and more, and more . . .

5

PRIMARY FOSSIL FUELS

Energy is produced from diverse fuels, each with distinct advantages and disadvantages and with its own social organization. There is no single, overarching energy industry. A somewhat oversimplified rule is fragmentation by primary fuel. One set of corporations and institutions is centered on coal, some owning electrical power plants that burn the coal. Another set is centered on petroleum. Since natural gas is often found with oil, there is considerable overlap of natural gas and oil companies, but there are organizations and networks that deal exclusively with natural gas. Electricity is not a primary energy source, but it too has a nexus of private and governmental organizations (see Part III), and within electrification there are specialty corporations and agencies involved with nuclear power and with hydropower. What is constant across the disparate fuels and technologies, and likely to remain so, is the motivation of their purveyors to increase the supply and consumption of their particular commodity, and for those organs that are profit-making to earn as much as they can.

Profit-making is not an essential feature of industrialization. The Soviet Union and its European satellites operated for decades through central planning, eschewing markets and profits as a means of allocating resources. The internal workings of communist societies were literally closed from the view of Westerners, who had little knowledge of how effectively central planning actually worked, or the impacts of communist industries on the environment. Because of this myopia, Westerners in the 1970s and 1980s often fingered capitalistic profitmaking as the prime cause of environmental depredation. After the collapse of the USSR in 1991, the full extent of pollution wrought by communist industry became abysmally visible, and it was no longer tenable to blame capitalism per se. Industrialization itself was the cause, whether capitalist or communist, made worse when pursued without limits or controls.

Today all industrial nations are mixed-market democracies and all espouse profitmaking. All have legal avenues for interest groups to influence government policy to the advantage of profit seekers, on the one hand, or to push for regulations protecting human health and the environment, on the other. Often these are conflicting goals, usually fought out in the courts, agencies, and legislatures, sometimes through media campaigns, and occasionally via popular protest movements. Excepting Russia, blatant corruption, including bribery and coercion, apparently does not play a major role when today's industrial nations adjust their energy policies (at least compared to Third World nations), but exerting influence through lobbying, personal connections, and campaign contributions is legal and routine. In that sense the industrial nations, certainly the United States, are plutocratic with influence skewed toward the rich and powerful.

This chapter describes the physical and social characteristics of the three fossil fuels: coal, oil, and natural gas. All have their origins in ancient forms of life and exist today in finite quantities. For each fuel, some is concentrated in easily exploitable deposits, while some is dispersed and will never be worth extracting. The boundary between extractable and non-extractable fuels is a moving one, determined by the technology that is available and the cost willingly paid to reach the more difficult deposits.

All three fossil fuels are important components of the world's energy mix. It must be said in their favor that all have very high energy content per unit of weight or volume. For example, one liter (0.26 gallons) of gasoline can do as much work as 50 hours of labor by a preindustrial farmer, or a few hours of labor by a strong draft horse; a liter of gas will carry four people in a Toyota Prius for nearly 20 kilometers (12 miles). On the down side, all fossils fuels produce CO_2 and other pollutants during combustion, with coal the worst offender and natural gas the least. Despite utopian schemes to eliminate all fossil fuel use in the near term, it is likely that all will remain important at least through the middle of this century because of their high energy content, sunken investment in their existing infrastructures, their relatively low cost compared to renewables, their availability and convenience, and because the organizations and institutions associated with each fuel will promote their continued use.

Coal

Coal was the first fuel of the industrial transformation, powering factories, railroads, and steamships. In the form of coke it is essential for the manufacture of steel. Its rapid growth in consumption caused William Stanley Jevons to warn in *The Coal Question* (1865) that Britain would soon run out. Today we see the major advantages of coal as its cheapness and abundance. (When government puts a price on CO_2 emissions through a carbon tax or some other device, coal loses some of its cost advantage.) There are enormous deposits in North America, Eurasia, and Australia. With relatively simple technologies, coal is easy to mine and transport, therefore inexpensive, and each lump contains a lot of energy (i.e., it has high energy density or "heat content"). It is also the most polluting source of energy and emits more CO_2 per unit of energy output than other fuels.

Until the mid-twentieth century coal was commonly used for space heating. I have vague memories of a coal truck making deliveries to the apartment building where I lived in Chicago, dumping its load down a chute into the basement, where our janitor would shovel it into a furnace that heated the entire building. A labor-saving technological advance of the day was a mechanical stoker that could be filled once and would automatically feed the furnace.

Coal mining in its early, unregulated days was arguably the most dangerous job on the planet, the workers threatened by mine collapse or flooding, underground gas explosions, and black lung disease from chronic exposure to coal dust. There were hazards of blasting underground coal seams and transporting loads to the surface. Children were employed in some of the most routine but dangerous work. Miners often lived in miserable communities where the entire economy was controlled by the mine owner, their wages barely covering the cost of food and other supplies purchased from the company store. Sons of miners became miners too, for lack of other opportunities.

The rise of trade unions improved working conditions in many fields, but probably none more importantly than the unionization of coal miners in the early twentieth century. Wages and safety were moderately improved, but at a cost of continual struggle between the unions and mining companies, often involving the threat or actuality of long strikes, which were punishing to both sides, as well as choking off the nation's coal supply. In Britain after World War II, unionized miners were regarded as so obstructionist that the coal industry was nationalized, not returning to private control until the administration of Margaret Thatcher.

The fall and return of coal

Even before the modern environmental movement forced industrial pollution onto national agendas, coal was obviously so dirty and dangerous an air pollutant that its use in cities was restricted after World War II. Oil and natural gas were by then available as cleaner, more efficient, and economical options for space and water heating. The railroads replaced coal-fueled locomotives with superior performing diesel engines. Railroad owners enjoyed the corollary benefit of eliminating the fireman on every locomotive, though not without union opposition. In North America, the switch from coal to diesel locomotives coincided with the general decline of railroads, encouraged by the automobile and truck industry and the construction of an interstate highway system. In Europe and Japan, where gasoline was more expensive and roads not as good, railroads were maintained and improved. Riding on a smoothly running high-speed train is for me the nicest way to travel across Europe.

By the 1960s coal had essentially vanished as a space heating fuel in industrial nations, in sharp contrast to China or India. The seemingly "obsolete" fuel was delightfully romanticized in Walt Disney's *Mary Poppins* (1964), showing Edwardian

London's coal-dusted sky (in which particulates, by increasing the refraction of red light, could produce splendid sunrises and sunsets) as background while Bert and his fellow chimneysweeps dance happily across the rooftops, oblivious to the scrotal cancer and other occupational hazards caused by carcinogenic soot.

The future of the American coal industry, which seemed so bleak in the 1960s, turned upward in the 1970s after spikes in oil (and natural gas) prices made petroleum too expensive for electrical utility companies in the US, so they converted oil-fired power plants to cheap and plentiful coal. Climate change had not yet been identified as a problem, so CO_2 emission was of no concern. Newly implemented air pollution regulations were an issue but not insurmountable. People were complaining about nuclear power but not about coal, which came roaring back. Automation reduced the labor force, and coal-smudged underground miners were largely replaced by operators of big machines. American railroading became heavily invested in the transport of coal, sometimes carrying Wyoming coal as far east as Massachusetts.

Trends in per capita coal consumption for the major industrial regions – the United States, the European Union (as it is constituted today), and Japan – are shown in Figure 5.1. European consumption (per capita) declined during the 1990s, partly because of the newly recognized problem of global warming but also because coal was not as cheap compared to other fuels as in the US. Japan, which has no indigenous coal, considerably increased its importation after 1980. All of the industrial regions showed drops in consumption following the Great Recession of 2008.

Chinese consumption is shown in Figure 5.1 for comparison. While per capita consumption of coal remains far higher in the US than elsewhere, the rapid rise in Chinese coal burning during the past decade is remarkable. Since China's population is roughly five times that of the United States, its total coal consumption is by far the highest in the world, and China now exceeds the US in CO_2 emissions.

The coal industry

"Big Coal" is the phrase used by writer Jeff Goodell (2006) for the alliance of coal mining companies, coal-burning electric utilities, railroads, lobbying groups, and industry supporters that make the coal industry such a political force in America. Unlike the giant oil companies whose operations span the globe, Big Coal's focus is usually domestic, though Japan and Germany import coal in large amounts for electric power generation. Also unlike petroleum, there is no need to find new deposits. Every industrial country rich in coal has a similar set of organizations to extract and convert coal into electricity, to increase consumption, and to influence government policy favorably.

The resurrected American coal industry strove for economic efficiency and, with regulatory enforcement, improved safety. Small mines were closed or consolidated into larger operations. Peabody Energy Corporation, headquarters in St. Louis, Missouri, is now the world's largest private-sector coal company. It has mines in

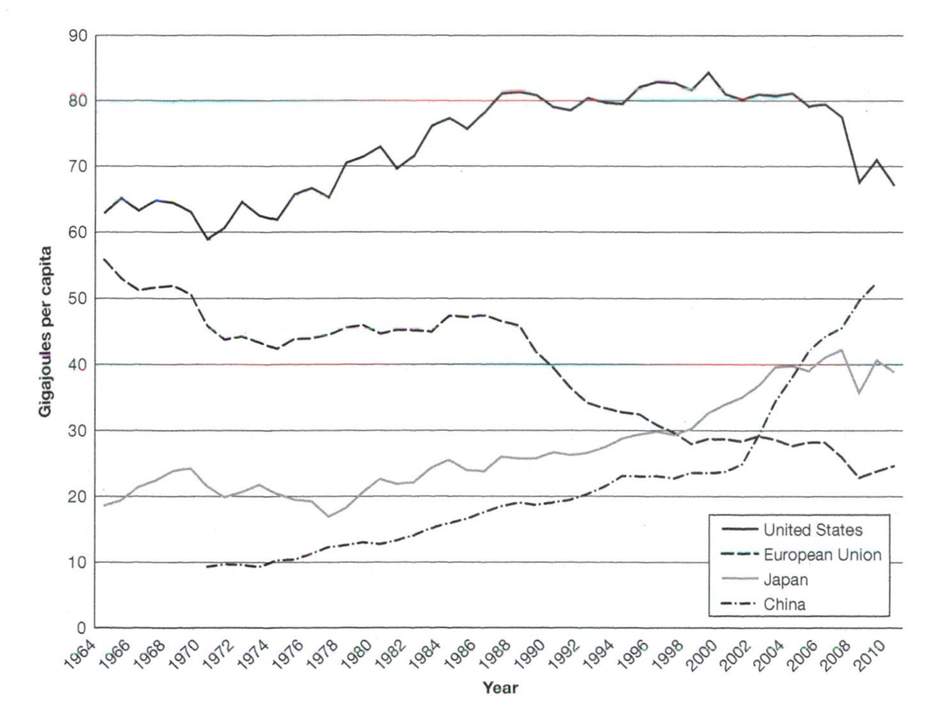

FIGURE 5.1 Coal consumption per capita for the United States, Europe and Japan, by year (BP, World Bank)

Australia and sells coal in many nations, but like other coal companies, most of its activity is at home where it operates surface and underground mines across the United States and produces ten percent of the nation's electricity.

"Coal" is a general term that includes a wide variety of solid fossil fuels. The broad categories most commonly used are anthracite, bituminous, subbituminous, and lignite, but there are many finer distinctions based on the purity of carbon (compared to other elements present), hardness, energy density, and visual appearance. Dull black bituminous is "garden variety" coal, its various grades composing the primary type first used during industrialization. Bituminous was and still is extracted from underground mines, but more recently its seams are reached by removing the overlay, including the mountaintop removal that has been much criticized in the American Appalachian region. Early railroads, burning bituminous, were far superior to wagons pulled by animals, but they puffed black smoke, an inconvenience for passengers hoping to arrive in clean clothes.

Anthracite is better quality coal, harder with higher energy content; it is glossy black, visually beautiful. But it is scarcer, often deeper, therefore more expensive to extract, and it requires high temperature to ignite. Almost pure carbon, anthracite burns more cleanly than bituminous. In early twentieth-century America, the Delaware, Lackawanna and Western Railroad began using anthracite coal from its mines in Pennsylvania, luring passengers with the promise that they would arrive at their destination in pristine condition. The railroad's advertiser created a character named Phoebe Snow, a young New York socialite who expressed her appreciation in a series of jingles, the first being:

> *Says Phoebe Snow, about to go, upon a trip to Buffalo,*
> *"My gown stays white, from morn to night, upon the Road of Anthracite."*

It was honest advertising, and Phoebe Snow became a widely recognized icon.

Subbituminous coal is softer and more crumbly than bituminous and has less energy density. Vast deposits in Wyoming and Montana have become desirable in recent decades because of their low sulfur content, so their combustion produces less sulfur dioxide than bituminous (but more CO_2) per unit of energy output, and this coal is inexpensive. Wide seams of subbituminous lay not far below ground surface and are easily extracted from open-pit mines; no underground shafts are required. Huge earthmoving machines first remove the overlay, then the coal, which is loaded on specialized trains that shuttle back and forth between mines and electrical power plants. Some power plants are built next to mines, eliminating transportation costs (but increasing the distance of electrical transmission). Miners in these new operations have good working conditions, operating their giant machines from comfortably enclosed air-conditioned compartments. In the US, subbituminous mines produce almost as much coal as bituminous mines (Goodell 2006).

Lignite, or soft brown coal, is the lowest quality that is commercially exploited, with the least carbon content, most impurities, lowest energy density, and highest CO_2 emission per unit of energy. Despite those disadvantages, it is used in countries where it is extremely plentiful and very cheap. Germany has the largest known deposits of lignite, those classified as economically mineable being greater than the entire oil and gas deposits of Europe (excluding Russia). One quarter of Germany's electricity generation is fueled by lignite, and that is expected to rise. This is a peculiar juxtapose to Germany's emphasis on reducing greenhouse gases through a very large increase in renewable energy production, while at the same time – as a reaction to Japan's Fukushima disaster – planning to close its considerable nuclear power capability by 2022 (RWE 2011).

German mining of lignite is remarkable to see. I visited the Hambach site, the deepest open pit mine, its bottom the lowest surface point in Europe, its area now 34 km^2 and expected to expand to 85 km^2 (http://en.wikipedia.org/wiki/Sophienh%C3%B6he). The mine is easily visible on Google Earth, just east of the small city of Jülich, Germany. Next to the pit is the Sophienhöhe; at 290 meters in height it is the largest artificial hill in the world, a pile of earth dug from the pit. With changing environmental attitudes, the hill is no longer growing because now extracted dirt is put back once the lignite is removed. To accomplish this, giant earthmovers crawl along the one side of the pit, scraping off overlay and excavating huge steps down to the level of the lignite. The lignite is then extracted and sent to nearby power plants. At the far side of the pit, more giant earthmovers replace dirt recently extracted. In effect the pit "moves" along the landscape, simultaneously excavated on one side and filled on the other.

Electricity from coal

Today, nearly all coal used by industrial nations is for the generation of electricity. Usually the coal is first pulverized to improve the efficiency of combustion. The fire boils water, producing steam that turns a turbine, a device for rotating a loop of wire in a magnetic field. This produces electric current in the wire, which is sent from the generation station into an electric grid, through which it reaches end users. Typical of steam generators, about one-third of the energy input to the plant (as coal) is output as electrical energy, with two-thirds dissipated as waste heat, often to nearby bodies or water or to the air through cooling towers.

Restrictions on emission of SO_2 are implemented either by using coal that has low sulfur content, or by installing "scrubbers" that remove SO_2 in the plant's smokestack. There are comparable methods for removing particulates. Pollutant reductions carry a monetary cost and reduce the efficiency of generation, so they tend to be resisted by coal burners. Often environmental regulators allow "grandfathering" of older plants so these can be maintained, modified, and remain operating despite relatively low efficiency or failure to meet environmental standards. This is a serious point of contention between Big Coal and environmentalists. In the US, the administration of President George W. Bush was more permissive than that of President Barak Obama. At the time of writing, with new restrictions on the older coal plants, plus the availability of cleaner and cheaper natural gas, aged coal plants are rapidly shutting down or converting to natural gas. In 2012 about one-third of US electricity came from coal, down from nearly half only five years earlier (Lipton 2012). It is too early to know if this is the beginning of a long-term trend.

Reduced consumption of coal by industrial nations may not translate into a global reduction in yearly emissions of CO_2. As coal producers lose their industrialized customers, market economics suggest that the price of already-cheap coal will drop further. If so, coal companies will find profit by selling their commodity to the Third World, as Australia now ships its abundant coal to China. (This seems like "shipping coal to Newcastle" because China has abundant coal of its own, but considering overland transportation costs, it is sometimes cheaper for the Chinese to import from Australia.) The net effect on global warming may be zero or even a worsening, since a lower price on coal is likely to produce increased consumption of the CO_2-spewing fuel in places where coal burning is not restricted.

The holy grail of the coal industry is to remove (or prevent) CO_2 as an emission, which is possible in theory but difficult in practice to implement economically on a large scale. Once CO_2 is removed, there is the problem of what to do with it, since it cannot be released to the atmosphere. One possibility is to pump it underground, but this carries uncertain risks of inadvertent release and destabilizing the ground.

Oil

Forbes magazine publishes yearly a list of the biggest companies, its ranking based on equal weighting of sales, profits, assets, and market value (http://www.forbes.com/sites/scottdecarlo/2012/04/18/the-worlds-biggest-companies/). Leading the list in 2012 was US-based ExxonMobil, which earned the largest single-year corporate profit ever: $41.1 billion. Another ten oil and gas companies ranked in Forbes's top 30: Royal Dutch Shell (Netherlands), PetroChina, Petrobras-Petrólea (Brasil), BP (Britain), Chevron (USA), Gazprom (Russia), Total (France), Sinopec-China Petroleum, ConocoPhillips (USA), and Eni (Italy). (Peabody Energy, the world's largest coal company, ranked 740th.)

Excepting Russia's Gazprom, a supplier of natural gas to Europe, all these huge international firms deal in both oil and natural gas. Many are vertically integrated, combining "upstream" exploration and extraction of petroleum with "downstream" refining and retail sales. Profits are mostly upstream (Coll 2012). A company that keeps a small fraction of the high price of each barrel of oil it sells will be highly profitable because it sells a lot of barrels. If price goes up and volume increases, profits will increase more, as long as expenses are kept in rein.

Trends in per capita oil consumption for the United States, the European Union, and Japan are shown in Figure 5.2. The effects of the oil shocks of the 1970s, especially the price increases following the Iranian Revolution of 1979, are obvious, though after oil prices declined in the early 1980s there was some rebound, especially in Japan until it entered a period of economic malaise in the late 1990s. As usual, per capita consumption in the US was and remains far higher than elsewhere. All the industrial regions show declines during the economic slowdown after 2008.

China is included in Figure 5.2 for comparison. Though its per capita consumption of oil is far lower than any of the industrial regions, the clearly increasing trend since the 1990s reflects the rapidly rising number of cars on Chinese roads that has accompanied urban China's new prosperity. Emulating Western habits of consumption, affluent Chinese now want larger homes and their own cars, even SUVs. Automobile manufacturing has become a major pillar of the Chinese economy, almost entirely based on licensing from Western corporations such as General Motors and Jeep. The often opaque air of China's cities is increasingly due to smog from automobile exhaust.

Global trade

More than any other fuel, oil is traded on global markets. It is easy to ship long distances across land via pipeline and across oceans via supertanker. Third World nations with the largest deposits of easy-to-extract petroleum do not use much themselves and are eager to sell; it is their major source of revenue. This income

FIGURE 5.2 Petroleum consumption per capita for the United States, Europe and Japan, by year (BP, World Bank)

may go into the pockets of a nation's governing elite, rather than raising the general welfare of its people, a phenomenon called "the resource curse" and rampant in Africa (Ross 2012).

When the gods distributed oil deposits around the Third World, they favored the Persian Gulf (Saudi Arabia, Iran, Iraq, Kuwait, United Arab Emirates, Qatar), Latin America (Venezuela, Brazil, Mexico), northern and western Africa (Libya, Nigeria), and the Caspian Sea (Kazakhstan, Azerbaijan) (BP 2011). This output is sold and prices set by traders on the world market. Exporting nations can influence price by making more or less oil available to the market, as is done in concert by the OPEC nations, which favor smoothly rising but not extraordinarily high prices that would encourage conservation and might rebound in retaliatory measures. Saudi Arabia is so large an exporter that it can, on its own, influence the world price, which it occasionally does in the interests of its protector, the United States, or to stabilize markets (Lippman 2012).

Figure 5.3 shows trends since 1960 of annual crude oil importation by the United States, Europe, and Japan. Populous Europe as a whole has always been the greatest importer, while Japan, despite having no indigenous oil, is relatively low in total imports because of its smaller population. Despite America's rich endowment of petroleum, it turned rapidly to imports in the mid-1970s because foreign oil was cheaper than the cost of producing domestic crude. After the oil price increase following the Iranian Revolution of 1979, all the industrial regions reduced their imports; this was also a time of economic decline in the West. When oil became cheaper during the mid-1980s (and economies recovered), they again raised their imports. In about 1991, Japan uniquely began a long-term economic slowdown, causing gradual decline in its oil imports. At the same time, Europe raised imports, and the United States raised them even faster. By the mid-2000s, the US was using more foreign oil than ever before, despite incessant but vacuous calls by politicians to reduce the America's dependence on foreign oil. Imports generally declined along with the economy after 2008.

There are large domestic deposits of oil in the industrial world – in Russia, the United States, and Canada. The United States uses its entire domestic supply, keeping large portions off limits for environmental reasons. Canada has so much oil for its small population that it has plenty to export but most is deposited as tar sands (bitumen) in the western province of Alberta. Extracting usable oil from tar sands is a messy, expensive process requiring a lot of water and energy, and emitting high amounts of CO_2, but with new technology and today's high price of oil it is going full bore. Possibly Alberta holds the third largest oil reserve in the world, after Saudi Arabia and Venezuela (Coll 2012).

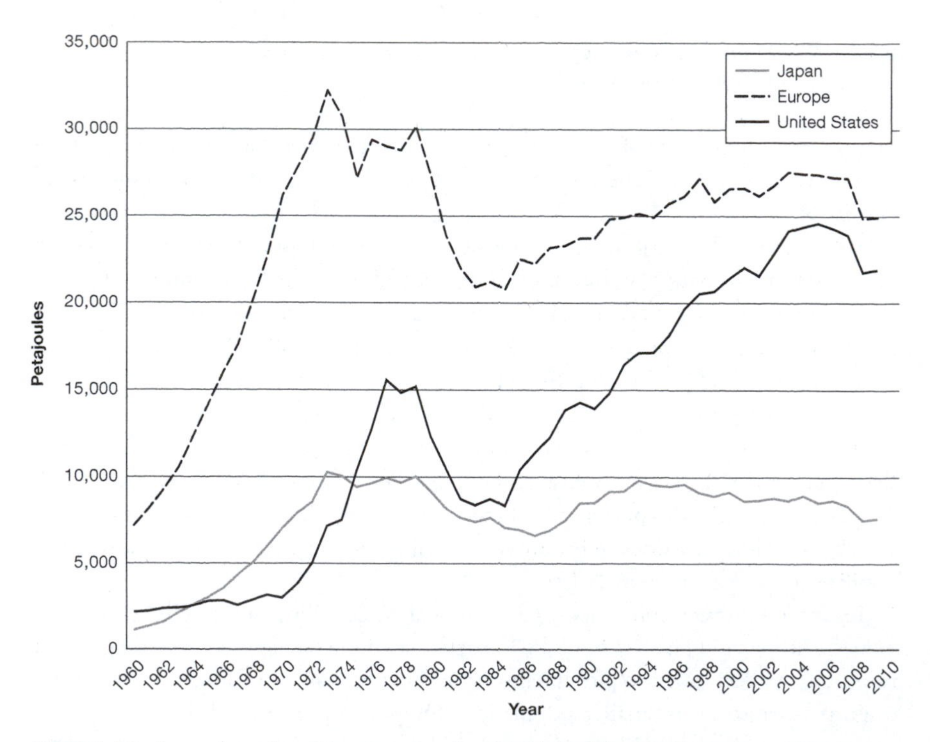

FIGURE 5.3 Annual crude oil importation by the United States, Europe and Japan (IEA)

Aggregate oil importation shown in Figure 5.3 does not distinguish the sources of import. These have changed since 1978, the earliest year when complete data on imports are available, and 2010. Figure 5.4 compares for these two years the major foreign nations sending crude oil to the United States (a), Europe (b), and Japan (c).

In 2010 Canada was the major oil exporter to the US, a considerable change from its unremarkable status in 1978. Larger shares of oil are also coming in from Mexico and Venezuela. These are sources close to home and thus relatively secure. The Chavez government of Venezuela is not rhetorically friendly to the US, but that does not interfere with business. On the other hand, Saudi Arabia, closely allied to the United States, is less a direct source of oil though the kingdom remains foremost in the interest of America to keep its oil flowing onto the world market. Generally speaking, the shift in imports reflects American security concerns and foreign policy. Iran and Libya, important in 1978, virtually disappear in 2010.

Europe brings in little oil from the Western Hemisphere, instead depending on the Middle East and former Asian republics of the Soviet Union. As a generalization, it has dispersed its sources, no longer depending as it did in 1978 on one or two major exporters like Saudi Arabia, though the Saudis remain the single largest foreign source of oil. Not shown here are exports internal to Europe, especially from the North Sea, a source exploited after the oil shocks of the 1970s.

Japan continues to depend very much on the Middle East, especially Saudi Arabia. Its biggest shift from 1978 to 2010 was replacing Indonesia with Qatar.

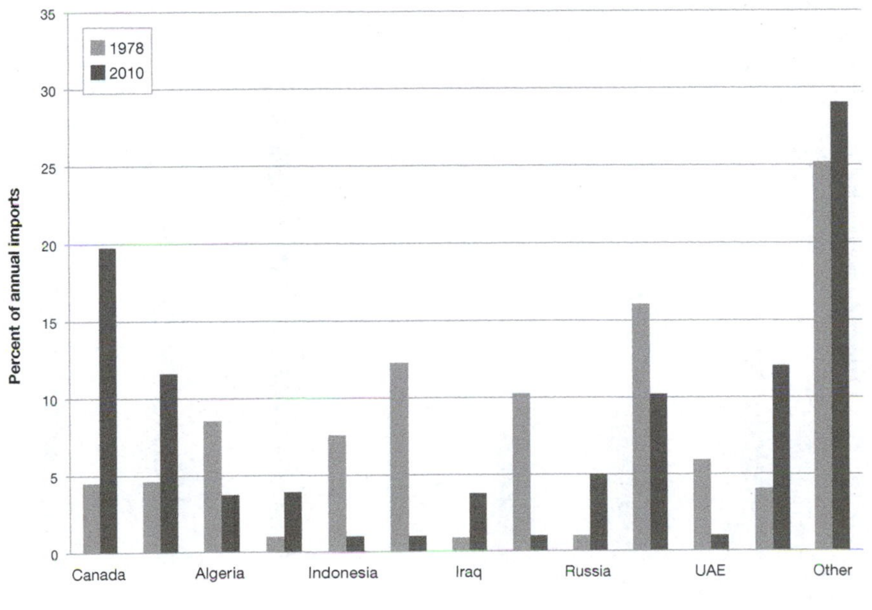

FIGURE 5.4A United States's sources of foreign oil, 1978 and 2010 (IEA)

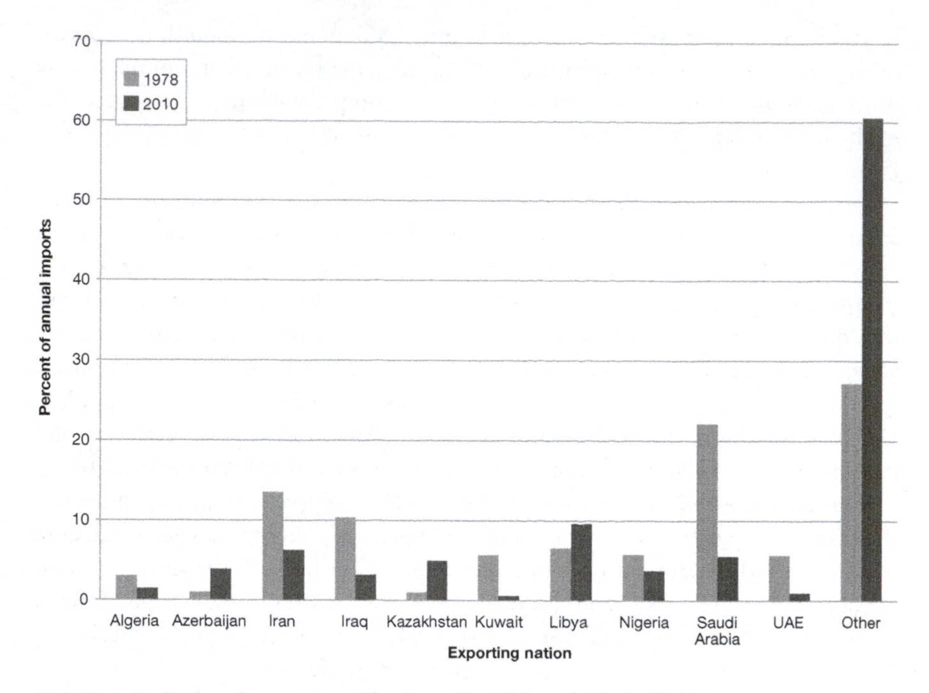

FIGURE 5.4B Europe's sources of foreign oil, 1978 and 2010 (IEA)

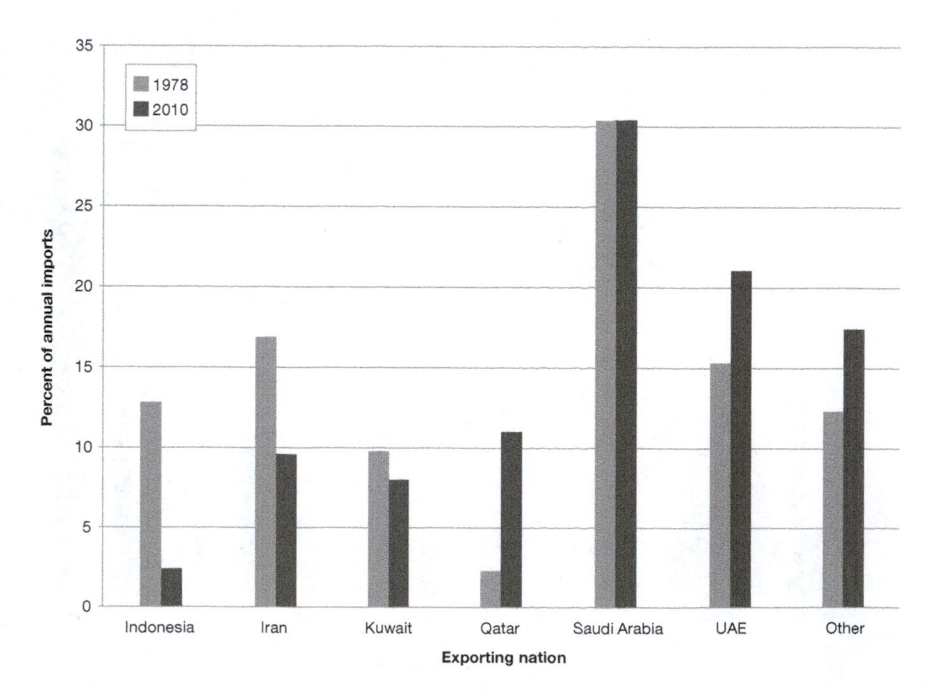

FIGURE 5.4C Japan's sources of foreign oil, 1978 and 2010 (IEA)

The often-heard goal of breaking a nation of its dependency on foreign oil is political rhetoric without important substance if the implication is to replace imported petroleum with domestic production. Oil, whether drilled at home or abroad, sells on a world market; it "flows" to the buyer offering the highest price. As a fungible commodity, it makes no economic difference to consumers where a barrel comes from; one is interchangeable with another. Ending the importation of oil would alter the international balance of payments, but by exploiting high-cost domestic sources in preference to low-cost foreign sources – a dubious advantage. Apart from economics, there are moral and strategic objections to sending petrodollars to authoritarian regimes where the money sometimes supports terrorists and often goes into the pockets of tyrants. This is not an objection to buying oil abroad, but to the character of the nation from which it is bought. In any case, if one nation no longer buys oil from Iran, other nations likely will (unless there are agreed-upon international sanctions that prohibit purchase of its oil). Canada, now the United States' major source of imported oil, is morally clean. There can be tactical problems when oil supplies are cut off by terrorism, warfare, accident, or embargo. This is not a matter of importation per se but of vulnerable supply lines, which might be more secure between Alberta and the lower 48 United States than between Alaska and the lower 48. The threat of an abrupt interruption of imports, like the OPEC embargo of 1973, is now lessened by creation of strategic oil reserves on domestic soil, held in case of emergency. America's oil problem is not that it imports too much but that it uses too much.

The changing oil industry

What President George W. Bush once called America's "addiction to oil" began innocently enough with the successful drilling in 1859 of an oil well near Titusville, Pennsylvania by Edwin Drake, who pumped oil out of the ground as you would pump water. By 1901 far more plentiful oil was gushing from Texas. The viscous black fluid is a mix of many different long-chain hydrocarbon molecules, each with different properties including temperature of evaporation. An oil refinery works like a bootlegger's still that makes moonshine out of fermented mash. The raw oil is heated, its different fractions vaporizing when reaching their characteristic temperature of evaporation, thus separating from the brew. Each vaporous fraction is then cooled, returning it to liquid state but now relatively pure. The brew is heated further and the next fraction taken off. With more processing, including the breaking (cracking) of very long molecules and removal of impurities, these become the refined products that today include gasoline, jet fuel, diesel oil, heavy fuel oil, etc.

By the end of the nineteenth century, John D. Rockefeller was the richest person in America, having built the Standard Oil Company's monopoly over production and sale of petroleum products. The growth of this behemoth corporation was not based on gasoline but on the sale of kerosene for lamps, which at the time was a good means of artificial lighting, a technological improvement over candles or whale oil lamps, and cheaper or more accessible than gas lights. When Thomas Edison began selling electric light bulbs toward the end of the century, demand for kerosene slackened, but that was when gasoline-driven conveyances were introduced, more than replacing lamps as the consumers of oil.

President Theodore Roosevelt took up "trust busting" as one of his favored domestic policies, his special target the unpopular Rockefeller's Standard Oil. This effort succeeded in breaking Standard Oil into smaller independent entities, the pieces eventually becoming Exxon, Mobil, Chevron, Conoco, and another that became the American branch of British Petroleum (later BP). Rockefeller owned shares in all the daughter companies, and as each grew larger, he became richer than ever. In the meantime or shortly after, petroleum combines such as Royal Dutch/Shell were growing in Europe, exploiting rich deposits in Russia, the Dutch East Indies, and Venezuela.

Oil was World War II's crucial fuel. During the postwar economic expansion, consumers and their automobiles caused an unexpected surge in demand for petroleum. Reaching for the plentiful, cheap, and easy-to-reach deposits of the Arabian Peninsula, American oil companies and the Saudi royal family formed a consortium, Aramco, the Arabian-American Oil Company. The small royalty that Americans paid the Saudis was a fortune compared to the kingdom's prior impoverishment, as was true in other oil-rich nations of the Third World, increasingly exploited by the West. A reliable flow of very cheap gasoline was fueling the Detroit-manufactured gas guzzlers of the 1960s as well as the rebuilding of war-ravaged Europe and Japan.

Major American and British oil corporations, with their joint international ventures and seemingly cartel-like control, were known derisively as the "Seven Sisters." These were (or would later be renamed) Exxon, Mobil, Chevron, Texaco, Gulf, Royal Dutch/Shell, and British Petroleum (later called BP), and there was an eighth sister if one counted France's Total (Yergin 1991). Behind the Iron Curtain, the Soviet Union was developing its own abundant deposits.

Led by Venezuela, several Third World oil exporters, mostly Arab nations, banded together in 1960 to become OPEC (the Organization of Petroleum Exporting Countries), but for more than a decade they still had no power over the sisters. OPEC's opportunity came with the Yom Kippur War of 1973, its Arab-led oil embargo and price hikes, which blindsided the importing nations. With the exporters now holding a strong hand, they demanded and got a larger share of profits and quickly assumed full control of their petroleum deposits. Some concessions that had been sold to the sisters would remain in place after 1973, but the termination of the last great ones – in Kuwait, Venezuela, and Saudi Arabia – marked the end of Western exploitation. The exporting nations formed their own national petroleum companies. They still needed the technical expertise of the West's oil industry, but now this would be acquired through business contracts between equal partners. It was an arrangement that would be highly profitable for both the exporters and the Western oil companies.

Natural gas

"With its natural gas and oil pipelines that tie Europe to Russia like an umbilical cord, Russia has unchecked powers and influence that in a real sense exceed the military power and influence it had in the Cold War" (Goldman 2010: 15). The Soviet Union could never use its nuclear missiles, knowing that the United States would retaliate in kind, but there is no deterrent against the world's largest producer of natural gas shutting off its pipelines. These are mostly owned by Gazprom (formerly the Soviet Ministry of the Gas Industry), ranked 15th on the *Forbes* 2012 list of the largest companies, its Moscow headquarters the central controller for gas flow from Russia to Europe. (After Russia, Norway and Algeria are the next largest exporters of natural gas to Europe.)

The Russian pipelines pass through Ukraine and Belarus, formerly republics of the USSR, now independent nations. Both sometimes refuse to pay the high price for gas demanded by Gazprom. Usually at such times, Gazprom cuts off part of the flow intended for sale in those two nations while demanding that they continue to pass the remaining flow to Russia's West European customers. In 2006 and 2009, Ukraine and Belarus not only refused to pay the higher price but also refused to pass along the gas, forcing the West Europeans to draw from their natural gas reservoirs.

Despite such vulnerabilities, natural gas is in many ways the most desirable of the fossil fuels. It is the least polluting and the lowest emitter of greenhouse gases (emitting about half the CO_2 of coal for the same energy output). Unburned methane is a worse greenhouse gas than CO_2, so this benign assessment of the fuel presumes that there is very little leakage from natural gas pipelines or production facilities.

The gas industry

Natural gas consists mostly of methane, the simplest hydrocarbon, but also contains more complex gaseous hydrocarbons (ethane, propane, and butane) and some impurities. These other components are extracted before the gas is sent to homes, businesses, and industry as nearly pure methane, a clean-burning gas that is colorless and odorless. (The odor of natural gas is added purposively to ease detection of leaks.) Tanks of propane are commonly seen in rural areas not served by pipelines.

The gas industry in America and Europe began not with natural gas but with manufactured gas made by heating coal in the absence of oxygen. By the late nineteenth century, thousands of companies were selling "coal gas," sending it to nearby customers through leaky (often wooden) pipes for lighting streetlamps, homes, and business. In cities with gas works, it was an alternative to kerosene lamps until both were replaced by electric lights. By that time, efforts by utility companies to promote coal gas for cooking and heating water were so successful that these were its most important uses (Busby 1999).

In the early days, natural gas coming out of oil wells was a nuisance to drillers. They couldn't easily transport it, so they burned (flared) it off on the spot. When huge amounts of natural gas were found along with oil in Texas and Louisiana, or near the Caspian Sea, long distance transport was not economically or technically practical until steel pipes with welded seams was introduced in the 1920s. In America, new companies specialized in pipeline construction and operation, eventually reaching from the Gulf States all the way to New England. These companies bought gas from the drillers, which were often oil companies selling it as a byproduct, and sold it at the other end of the pipeline to local utilities or industrial customers. After World War II the construction of long-distance pipelines boomed; by 1950 America had more mileage for natural gas than for oil.

Steel (more recently plastic) pipelines allow gas to be raised (by compressors) to high pressure so that more of it is transmitted without serious leakage. As the pipeline network grew, so did the need for safe, usually underground storage facilities, located as near as possible to gas markets. Pipeline flow is smoothed out by storing gas in the summer months when demand is low, and withdrawing it from storage during the winter heating season. Depleted gas fields are the most prevalent media of storage. Newly piped gas is injected down old gas wells, deposited under pressure until needed. Aquifers, caverns, and abandoned coal mines are other options for storage.

In America during the postwar decades, rising demand relative to the supply of natural gas brought high prices, but sometimes oversupply brought low prices. Pipeline companies sometimes had to pay for gas they didn't need but were committed by long-term contracts to buy, and sometimes they had insufficient gas to transport. Seeking to stabilize prices at a reasonable level, the American government set limits on what producers could charge for gas at the wellhead. It was not a good solution. Drillers lost their incentive to find new deposits or exploit known ones, so supply lagged demand and prices rose, which in turn reduced demand. In 1989 the Congress removed the limits on prices. Seeing potential profit, the producers responded by increasing supply, while distributors encouraged their customers to consume more. Another result of deregulation was the proliferation of gas marketing companies. The most famous, Enron Corporation, was a leader in making gas a marketable commodity, but its executives and traders played fast, loose, and illegally with the market until the company went bankrupt and its president went to jail.

Figure 5.5 shows trends in per capita consumption of natural gas for the United States, the European Union, and Japan. (China is again included for comparison, and though we see an increase in the past decade, its per capita consumption of natural gas remains very much lower than in any industrial region.) It will be no surprise that per capita consumption in the US was and remains far higher than elsewhere. What may be surprising are the very high levels of American per capita use in the late 1960s, then the sharp dip, then the recovery in the 1990s. This erratic course, in contrast to the smoothly rising trends in the EU and Japan, is due to the changes in US government regulation of natural gas prices: first price limits, then in 1989 deregulation.

Modern pipeline networks are analogous to electric grids in having long-distance conduits of high-but-limited capacity that connect energy producers or marketers to local substations, from which the gas/electricity is distributed to

FIGURE 5.5 Natural gas consumption per capita for the United States, Europe, Japan, and China by year (BP, World Bank)

individual customers. (The most important difference is that electricity cannot be amply stored so at every instant, electricity supply must match electricity demand.) Often the same company that distributes natural gas to customers in its region distributes electricity too. Customers pay the company a monthly bill roughly determined by the amount of gas and electricity consumed as measured on their gas and electric meters.

The consolidation of "gas and electricity" utilities is best explained historically as a merger of "natural monopoly" companies, but there is a logic to it because of similarities in management problems. Their expenses, charges to customers, and profits are to varying degrees regulated by government; American states each have a public utility commission handling that. They must apply to government agencies in what is often a complex process of getting permits for new high-voltage electric transmission lines and high-capacity gas pipelines. Often these are opposed by citizen or environmental groups, elevating the importance of public relations. There used to be the problem of landowners blocking a proposed route by demanding exorbitant payment for crossing their land, but usually this is legally barred because utility companies can invoke the "right of eminent domain," which allows them rights of way through private land for fair compensation to the landowner.

Natural gas in the future: LNG and fracking

Since natural gas is usually found with oil, the areas of the world rich in oil have large deposits of gas too. With concessions in these parts of the Third World, the giant oil companies ExxonMobil and BP are now among the largest producers of natural gas.

Much of ExxonMobil's gas production is in partnership with Qatar Petroleum, the state-owned gas and oil company of the small Arab monarchy on the Persian Gulf. Qatar is unusual, for its enormous natural gas deposits are unmixed with petroleum. These were not exploited for some time after discovery because the gas was "stranded," sitting thousands of kilometers from customers who wanted it, but too far for pipelines to be economical. The eventual solution was to cool the gas, condensing it into liquefied natural gas (LNG), which is transported on refrigerated container ships. On arrival the LNG is unloaded at a receiving terminal, warmed to return it to its gaseous state, then put into pipelines for overland distribution. Some opponents of LNG terminals envision accidents at these ports in which a refrigerated container is breached by terrorism or accident, its contents spilling into the air and gasifying, causing a cloud of methane to spread over harbor and city until ignited by a chance spark, causing a conflagration. No such catastrophe has occurred, and whether or not it is a real risk is debated.

In the 1990s the United States was preparing for increased importation of LNG when the game-changing technology of hydraulic fracturing ("fracking") sent domestic production of natural gas skyward. At the time of writing, North American production is so high that the price has crashed. Now the process is being implemented elsewhere, perhaps beginning what the International Energy Agency calls a "golden age of gas" (IEA 2012).

The natural gas released by fracking is tightly situated in deep strata of shale rock, formerly too difficult and costly to extract. Fracking combines two techniques, horizontal drilling and fracturing of the shale to release the gas. The drilling begins with a normal vertical shaft, reaching the targeted shale stratum a kilometer or more under the surface. At that depth, the drill turns at a right angle, boring horizontally into the stratum. Several horizontal holes eventually radiate from a single drilling platform, each reaching a kilometer or more from the vertical shaft, opening multiple channels into the shale. Drillers slowly force water under high pressure into the well. This pressure reopens old sealed cracks (fractures) in the shale and opens new ones. Freed from the impermeable shale by the fracturing, the gas flows up the shaft. Water is used in large volume and contains grit to hold open the cracks, and chemicals – some toxic when highly concentrated – to lubricate and otherwise facilitate the process. Fracking has been successfully applied to "tight" shale and sand oil deposits too.

The process is controversial (Davis & Hoffer 2012). Especially in Pennsylvania and New York State, which sit atop the gas-rich Marcellus Shale, opponents have mobilized to stop fracking. Residents near some fracking wells have complained of toxic chemicals in their water, poisoned farm animals, and perhaps sickened people. Some residents are able to put a match to their faucets, setting the running water aflame because of natural gas coming out of the tap. Gas producers rejoin that water in the area has naturally occurring methane and that the flaming phenomenon has been known for years before fracking began. Some fear that fracking destabilizes the land, potentially causing earthquakes. This last is unlikely except in extraordinary conditions because the pressure of the frack water and size of the fractures are very small by geological standards. It is also improbable that injected chemicals can seep from the shale up to the water table because these levels are separated by a kilometer or two of fairly impermeable rock. Another claim is that the large volumes of toxic water that come out of frack wells have been flushed into streams; if true, the effects on water quality and ecosystems remains disputed. Drillers are beginning to partially recycle the water.

There is a vulnerability where the vertical shaft runs down through the water table. Drillers place a concrete sleeve around the drill hole at that level to prevent any of the chemical-laced water, or the outflowing gas, to make contact with the water table. But sometimes these cement sleeves are installed improperly, allowing seepage of gas or water from the well shaft to shallow aquifers near the water table. An undeniable social problem is that wholesale fracking brings truck traffic and drilling sites to rural areas that in some cases were bucolic.

On the positive side, rural landowners, often people of little means, have received payments of hundreds of thousands of dollars for leasing their land to gas drillers. A preliminary assessment of exploitable gas shale in 32 nations reports its widespread distribution, especially abundant not only in North America but also in China, Argentina, South Africa, and Australia. Among European nations that presently import natural gas from Russia, France and Poland have large shale formations (EIA 2011).

Conclusion

Present reliance on fossil fuels is so great, and so many powerful organizations promote their continued if not growing consumption, that it is hard to see how fossil fuels can be eliminated by mid-century even in industrial nations, much less in China and India. If American coal were no longer consumed domestically, American coal companies would sell their abundant supply abroad. Oil has stronger corporate and national promoters, and a ready market in the Third World.

To reduce the buildup of greenhouse gases in the atmosphere, and collaterally reduce other problems caused by fossil fuels, it seems unavoidable that the most energy-consuming nations must agree to limit their use. The goal need not be zero, but certainly a considerable reduction worldwide, with most of the onus on industrial nations that have gotten rich by burning these fuels for two centuries, and on China, which now leads the world in coal burning and CO_2 emissions. This implies limits not only on the fuels, but on the organizations promoting their consumption, probably in the form of monetary disincentives and regulatory constraints on production and sales.

Natural gas, the most environmentally benign of the fossil fuels, has been proposed as a "bridge fuel" during a decades-long transition period when coal and oil are replaced by some combination of renewables, nuclear power, and conservation through improved efficiency. This is a technically if not sociologically feasible strategy because new technologies like fracking, though unpopular, have made a considerable quantity of natural gas accessible at reasonable cost.

6

NON-CARBON SOURCES OF ENERGY

There is nearly consensus among industrial nations that we *should* reduce our dependence on hydrocarbon fuels. There is vast disagreement on how quickly or completely to do this, the means of doing it, and the amount of money that is justifiably spent to do it. The magnitude of transformation can be seen at a glance in Figure 6.1, showing the mix of primary energy sources used since 1965 by the United States (a), the European Union as presently constituted (b), and by Japan (c). Each industrial region has been and remains enormously dependent on fossil fuel, mostly oil. The only non-carbon sources of any significance are nuclear power and hydroelectricity. Other renewables have come onto the scene since the 1990s, but still they barely change the overall picture. Three categories of non-carbon energy are described here: nuclear power, renewables, and geothermal power.

Nuclear power

Mindful of public fear over the nuclear weapons buildup in the United States and Soviet Union, President Dwight Eisenhower proposed in 1953 his "Atoms for Peace" program. The idea was to shape nuclear energy into a constructive force for society by using it to produce unlimited electricity without air or water pollution (and most importantly for today's proponents, without greenhouse gases). The United States would help other nations build research and power reactors, but only if they were open to inspection by the International Atomic Energy Agency, to prevent plutonium produced in reactors from being diverted to make bombs. Thus, Atoms for Peace had dual roles: improving America's image, and controlling the proliferation of nuclear weapons. The British and Soviets were also formulating plans to export power reactors. In these early optimistic days, nuclear power was promoted by Lewis Strauss, then head of the US Atomic Energy

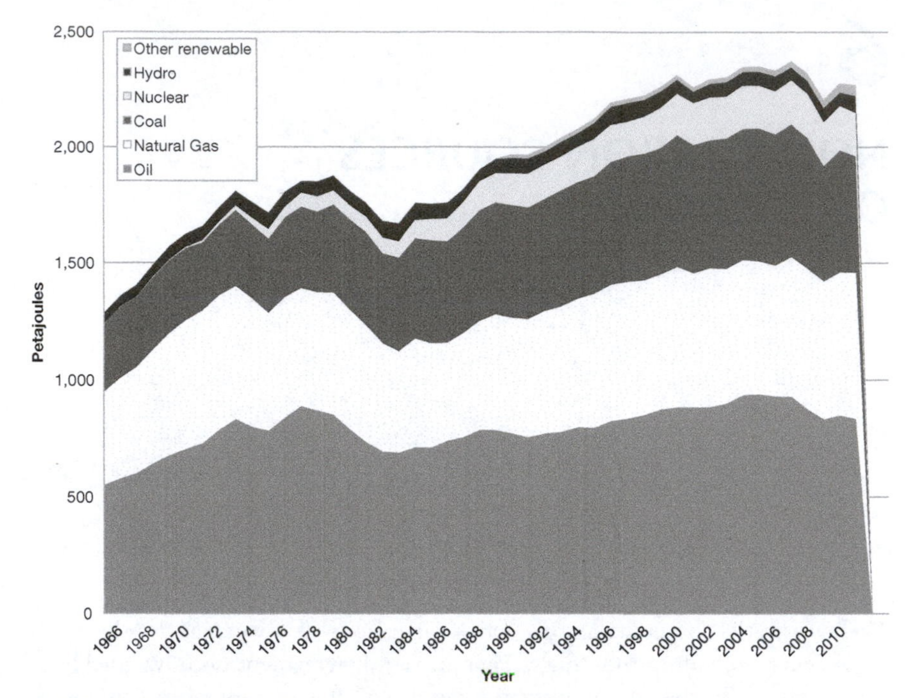

FIGURE 6.1A United States: Primary energy sources by year (BP)

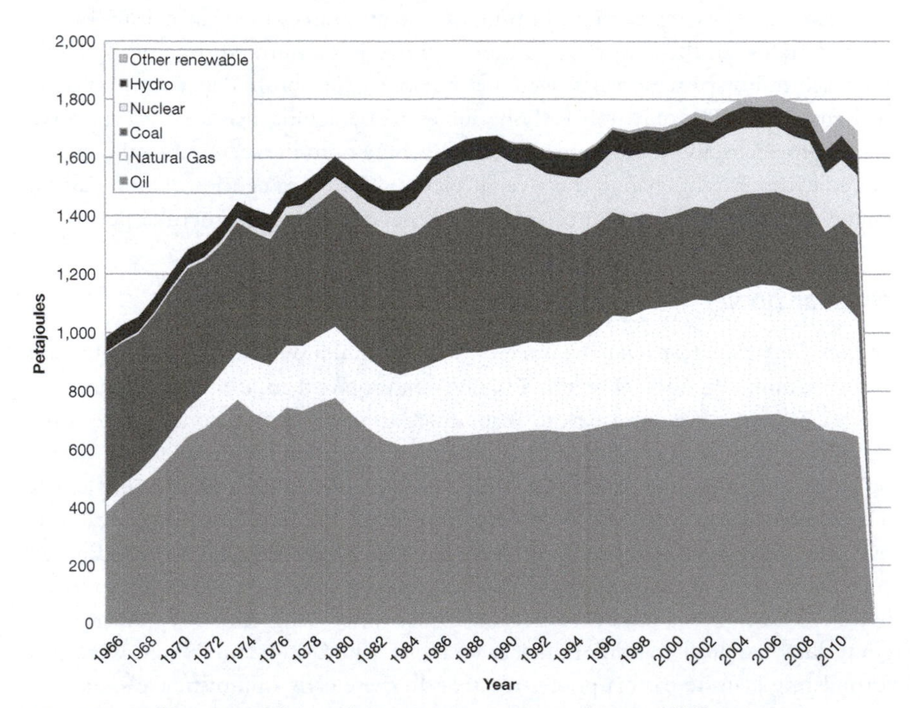

FIGURE 6.1B European Union: Primary energy sources by year (BP)

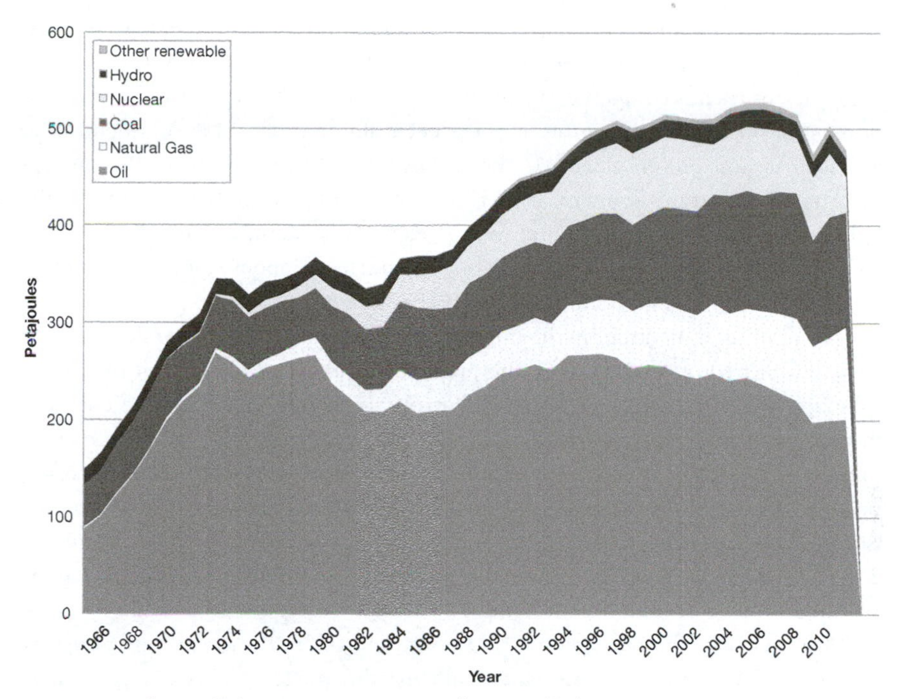

FIGURE 6.1C Japan: Primary energy sources by year (BP)

Commission, as "too cheap to meter" (Weart 2012: 85), a prediction that would turn out to be wildly far from reality.

Nuclear power plants are children of the US Navy's nuclear-powered submarines. Sub reactors were easily adapted to nonmilitary power generation and in slightly different versions became the heat source for most civilian nuclear plants built by Western manufacturers. (Soviet power reactors were descended from ones built for producing military plutonium.)

A nuclear plant is like a fossil-fueled power plant in one important sense: A source of heat, whether burning coal or a fission reactor, turns water into steam that drives a turbine that spins an electric generator. All thermal electric power plants work this way. The devil, or as some have said, "the Faustian bargain," is in the details.

Bombs

Nuclear fission, the splitting of heavy nuclei of uranium or plutonium atoms, provides the enormous power for atomic bombs. Fission must not be confused with nuclear *fusion*, the welding together of nuclei of light elements like hydrogen or lithium, which produces the far greater power of hydrogen bombs and energizes the sun. Fusion's prospects as a controlled source of civilian energy are too distant in the future to consider here.

Consider a simple "solar system" model of an atom. *Electrons* – very light particles with negative charge – orbit around a heavy nucleus, like planets orbiting the sun. The nucleus itself is composed of *protons* (heavy particles with positive charge) and *neutrons* (a bit heavier than protons and electrically neutral; think of a neutron as an electron and proton glued together). The number of protons in the nucleus determines the atom's element. If there is one proton, the atom is hydrogen, the lightest element; if there are 92 protons, the element is uranium, one of the heaviest elements. The number or electrons orbiting the nucleus is about equal to the number of protons, their negative and positive charges nearly in balance.

The number of neutrons in the nucleus can vary without changing the element. An atom of hydrogen always has one proton, but different versions of hydrogen, called *isotopes*, have zero, one, or two neutrons. This does not affect the chemical properties of hydrogen, but it does change the weight of the nucleus. Over 99 percent of naturally occurring uranium is an isotope with 146 neutrons. Since protons and neutrons weigh nearly the same, the total "atomic weight" of this isotope's nucleus is 92 + 146 = 238, and the isotope is denoted ^{238}U or uranium-238. Less than one percent of naturally occurring uranium is an isotope having 143 neutrons, its atomic weight being 92 + 143 = 235.

The nucleus of this rare isotope, ^{235}U or uranium-235, has the peculiar property that if hit by a stray slow-moving neutron, the nucleus splits nearly in half, converting the uranium-235 atom into two smaller nuclei of lighter elements, called "fission products." (The most common fission products are iodine, cesium, strontium, xenon, and barium.) This is literally the transmutation of one element into another, a process sought for centuries by alchemists who hoped to transmute lead into gold. When a ^{235}U atom fissions, it also releases a remarkably large amount of energy from the splitting nucleus, far more than the chemical energy released by oxidizing a molecule of hydrocarbon. During World War II, physicists realized that if they could simultaneously hit a large number of ^{235}U nuclei with slow moving neutrons, they could produce a very big explosion.

As it happens, when a nucleus of ^{235}U splits, it not only produces two new elements and excess energy, but it also throws off three neutrons. Usually these neutrons fly off into space. But physicists understood that if they packed atoms of ^{235}U closely together, they increased the probability that one of these escaping neutrons would hit another ^{235}U nucleus, which in turn would split and throw off three of its own neutrons, and so on. By packing the ^{235}U atoms closely enough, one could achieve a self-sustaining chain reaction, a cascade of ^{235}U fissions that could instantly blow up a city.

Since ^{235}U is less than one percent of naturally occurring uranium, a far lower concentration than is necessary to sustain a chain reaction, the builders of atomic bombs had to "enrich" their uranium, raising the portion that is ^{235}U to over 85 percent, a very difficult process because there is no chemical difference between ^{235}U and ^{238}U. (Separation usually requires sophisticated and expensive centrifuges that take advantage of the different nuclear weights of the two isotopes.)

A nuclear power plant cannot blow up like an atomic bomb, even in the worst conceivable accident. Uranium fuel for power plants is enriched sufficiently for a sustained release of heat to boil water, but far less than is required to support a nuclear explosion.

Before we leave the topic of bombs, there is one more matter of importance. The uranium fuel rods in a nuclear reactor are composed mostly of the common, non-fissionable isotope ^{238}U. While the reactor's chain reaction is running, a stray neutron often hits the nucleus of a passive ^{238}U atom. The neutron is absorbed into the nucleus, immediately forming a new isotope, ^{239}U. This is an unstable condition for the nucleus, which very soon throws off an electron. Remember that a neutron is like a proton and electron glued together, so the departure of that electron from the nucleus effectively converts the newly acquired neutron into a newly acquired proton, i.e., it raises the number of protons from 92 to 93. Thus, the uranium atom transmutes into a new element, plutonium-239, not found in nature. Like ^{235}U, plutonium-239 can be used to make an atomic bomb. (One of the A-bombs dropped on Japan was built of uranium, the other of plutonium.) Since plutonium is chemically different than uranium, it can be separated out from used fuel rods by methods commonly used in the chemical industry, within the capability of rogue states or terrorist groups. It is not as easy to detonate a plutonium bomb as a uranium bomb, but the prospect is still worrisome and one of the major demerits of nuclear power reactors. On the positive side, the plutonium produced in a reactor can itself be used to fuel other nuclear reactors, so it actually expands the supply of exploitable fuel.

Power reactors

Fuel rods at the core of a nuclear power reactor contain modestly enriched uranium. The rate of the chain reaction is controlled by water surrounding the uranium rods and by the insertion or removal or boron rods that absorb neutrons. (When boron rods absorb all stray neutrons, the chain reaction stops.) While the reactor is running, heat is transferred from the fuel rods to the surrounding water. This transfer serves first to cool the rods, which might otherwise heat to the melting point, and second to boil the water into steam. (In some designs, the hot water does not itself become steam but heats a second fluid in a closed loop into steam.) We then have the familiar sequence for a thermal electric plant: the steam turns a turbine, which is connected to a generator that spins a loop of wire in a magnetic field, producing current in the wire.

The uranium in new fuel rods emits little radioactivity and is not dangerous, but the fission products formed while the reactor is running are highly radioactive, some remaining so for nearly a million years. When a reactor is refueled, perhaps once a year, the spent fuel rods must be sequestered in a way that isolates them from people and ecosystems for millennia. During the early optimism about nuclear energy, it was assumed that spent fuel could be safely buried in stable geological formations. In retrospect, this was naïve, making no allowance for the difficulty of proving that a geological formation would remain stable for millennia into the future, or for public opposition to long-term waste repositories.

Reactor designers were cognizant of the possibility of accidents. Knowing that a reactor cannot explode like a nuclear bomb, they posited that the worst possible accident would be a loss of coolant to the reactor core, perhaps from a burst water pipe caused by an earthquake. Without coolant, fission products in the fuel rods would continue to heat, eventually melting the rods into a puddle of molten uranium on the floor of the reactor vessel. The puddle, still generating heat, might burn through the underlying concrete, then farther into the ground. Designers joked that molten fuel would melt all the way to China, hence "the China syndrome" as a nickname for the worst feasible accident.

More seriously, they feared that if molten fuel breached the containment, it might reach underground water, causing a ferocious steam explosion that would spew radioactive material into the air and across the countryside. To prevent this, engineers built redundant safety systems, including reactor shut-offs that automatically inserted boron rods to stop the reaction, emergency core cooling systems, and reinforced concrete containment (except in the Soviet Union) to prevent any radiation from escaping the site. Electric utility companies and reactor manufacturers were so confident of these safety features that one plant was proposed (though never built) in the midst of metropolitan New York, and the Indian Point nuclear complex was constructed and still operates on the Hudson River, less than an hour's drive from New York City.

Public opposition

The first public intervention against a civilian nuclear power plant was in 1956, aimed at Detroit's Enrico Fermi station and arising from a labor union dispute. The following year England's Windscale reactor, built to produce plutonium for bombs, suffered a fire and radiation release, provoking public outcry. These were isolated events but already one could see that activists tended to be on the political left, a circumstance possibly arising from union participation in Detroit and leftist opposition to nuclear weapons. In any case, this correlation between political liberalism and opposition to nuclear power persists to the present.

In the early 1960s there was a small cluster of plant interventions in the United States, most visibly at Bodega, California where a nuclear plant was proposed on the coastline. During initial excavations for the plant, an earthquake fault was discovered running through the site. It was further determined that there had been no movement along the fault for 40,000 years. Proponents of the plant claimed that the fault was inactive, and there was little likelihood of future movement. Opponents claimed that since there has been no earthquake for a long time, one was due. The Bodega site was eventually abandoned.

There was little protest in the mid-1960s, but in 1968, in the midst of high liberal activism over civil rights and Vietnam, local groups began to intervene against most nuclear plants that had been proposed in the US. News and periodical coverage increased to report these disputes and to carry propaganda put out by both sides. Opinion polls, by then appearing yearly, show a peak of opposition among the wider public in 1970, following a peak of mass media coverage in 1969. Then protest activity and media coverage declined into 1973.

Antinuclear activists showed new strength after 1973, accompanied by increased news coverage and higher public opposition as measured in opinion polls. The timing seems peculiar because this was the height of the "energy crisis" set off by the Yom Kippur War of 1973, with its worries about the adequacy of energy supplies. Yet here was rising protest against an alternate source of energy. A conundrum at the time, the explanation is now fairly well recognized. Briefly, unpaid partisans must allocate their limited time and resources between protest activity and other demands of family and work. They are most highly motivated toward protest activity when their issue of concern is high in the public mind, i.e., when it is highly covered in the mass media. Thus, continuous news coverage of the energy crisis during the 1970s actually reinforced the incipient movement against nuclear power, eventually spreading it around the industrial world.

In 1975 at Wyhl on the German side of the Rhine, 20,000 people gathered to protest a proposed reactor, tearing down the fence surrounding the site. Over the following months, activists from across Western Europe joined the encampment in an expression of counterculture with teach-ins, flowers, and singing. The Wyhl reactor was never built (Weart 2012). In France in 1977, 50,000 protesters marched on a proposed reactor site and clashed violently with police, but this time to no avail. In 1979, the accident at Three Mile Island, hugely covered in the international news, spurred more antinuclear protest and public dismay (Mazur 1981; Freudenburg & Rosa 1984). New political parties, the Greens, gained strength across Europe (though never attained much influence in North America). German Greens were very successful, eventually joining the government and winning Germany's avowal to end its reliance on nuclear power.

By the early 1980s, during the initially bellicose presidency of Ronald Reagan, many antinuclear power activists had shifted their time and energy to a more pressing problem, the reduction of nuclear weapons, joining the "Nuclear Freeze" movement. Protests against power plants were at low ebb and might have stayed there except for the Chernobyl disaster in 1986, one of the last spasms of the Soviet Union. Soviet plants did not have reinforced containment around their reactors, as Western plants did, to keep radiation from spreading offsite in case of a release. With no protective shell, the Chernobyl explosion that tore open the reactor building spewed radioactive material into the air, across Ukraine and Europe. Initially the Soviets did not acknowledge the accident, but radiation detectors across Europe told the story. The government later admitted that 31 people died of radiation sickness in the following weeks. Estimates of the eventual number of deaths from cancer are uncertain but are usually estimated in the low thousands. Because of high radioactivity near the site, an "exclusion zone" 60 kilometers in diameter remains in place, and hundreds of thousands of people were resettled.

The industrial nations reacted to the Chernobyl accident in diverse ways. France continued to build nuclear power plants, which now supply 80 percent of French electricity, as did Japan, eventually receiving a third of its electricity from the atom. Germany and Sweden imposed moratoriums on nuclear plants but later relaxed that position. In the United States there was no formal halt, but public opposition and high costs had essentially ended new construction after Three Mile Island, and Chernobyl intensified public aversion to nuclear power plants. All the while in the US, coal, which had been suffering long decline as a dirty, polluting fuel for space heating and industry, was resurrected without complaint as the cheapest and most plentiful fuel for electric generators.

After the response to Chernobyl dissipated, there were two decades without major accidents or protest, when public opinion toward nuclear power moderated, especially as its carbon-free nature was emphasized (Bisconti 2011). In the US there were quiet proposals to build a few new nuclear plants in areas of the country less likely to evoke new protests. That ended in 2011 when an earthquake and tsunami caused core meltdowns in three reactors at Japan's *Fukushima Daiichi* nuclear complex. Japan, until then one of the leaders in nuclear power, shut down all 50 of its reactors. Germany announced it would decommission all its power reactors by 2022. A year later, over tremendous public opposition, the Japanese government ordered a few reactors back online for fear that the nation's electric grid would otherwise be incapable of supplying the summer demand for electricity.

Nuclear waste

In 1987 the US Congress chose Yucca Mountain, Nevada, about 100 miles north of Las Vegas, as the site for America's permanent repository for long-lived highly radioactive waste from civilian power plants. Yucca Mountain is an extremely dry place and apparently stable geologically, but subsequent study showed that water does penetrate the mountain more than scientists initially estimated. In the meantime, local antinuclear opposition rose against the repository, and Nevada's Democratic senator Harry Reid joined the opponents. Complicating matters, in 2004 a federal appeals court threw out a set of federal rules for the site because they only required assured protection for 10,000 years, while the fuel would be hazardous for close to a million years. This strengthened the case of the opponents because it is unlikely that any scientific evaluation could assure the integrity of any geological formation for a million years (i.e., 100 times longer than humans have lived an agrarian mode of life).

By 2008, when Democrat Barack Obama was elected to the US presidency, Nevada's Senator Reid had ascended to Senate Majority Leader. The two quickly agreed to remove Yucca Mountain from consideration as a permanent repository, probably to cement Reid's essential support for Obama's legislative agenda. With no other candidate for a permanent repository, American power reactors – scattered across that nation – would each continue to hold its highly radioactive waste on site. Hardly any expert thought this a good arrangement but technical uncertainties and politics prevented the implementation of the preferred single repository. Possibly Yucca Mountain will reemerge as a storage site when President Obama or Senator Reid are no longer in a blocking position, or when a court order reopens work on that site. No other nation currently operates a large-capacity permanent repository for high-level waste from power plants. Sweden may be the first to achieve this.

Barely noticed in the public controversy over Yucca Mountain is that the United States Department of Energy (DOE) already operates a permanent repository for long-lived radioactive military waste from its nuclear weapons program. Located deep in a New Mexico salt formation, not far from the Los Alamos weapons laboratory, the Waste Isolation Pilot Plant (WIPP) was licensed to dispose of transuranic (i.e., involving elements heavier than uranium) radioactive waste for 10,000 years. There was public opposition to WIPP, but it never escalated to the degree of Yucca Mountain or gained as much political reinforcement. Despite inevitable uncertainties in the long-term integrity of the salt formation, WIPP began storing military waste in 1999. The DOE, working with linguists, anthropologists, and science fiction writers, intends to etch symbolic warnings in granite pillars atop the site, readable by future inhabitants, long after English-speaking Americans have passed from the scene (http://en.wikipedia.org/wiki/Waste_Isolation_Pilot_Plant).

Recent developments

In 2012, the United States had 104 operating nuclear reactors that produced about 20 percent of the kilowatt-hours used. Without any newly constructed plants since the 1970s, this was accomplished by older plants running longer than they had previously, now producing 90 percent of the electricity that they could theoretically produce if they ran continuously at maximum power. (There are necessarily shutdowns for maintenance and refueling.) Most of these plants still operate more or less like reactors on nuclear submarines. Newer designs, perhaps one day implemented, reduce dependency on active safety systems in favor of passive forces, for example, depending on gravity and natural heat circulation rather than pumps and valves to keep the reactor core supplied with coolant, and reducing the complexity of the equipment. The US Nuclear Regulatory Commission is considering approving only a few standardized designs, rather than in the past, approving each plant's design individually after construction had begun. This could cut the current ten-year construction process in half.

Prior to Japan's Fukushima Daiichi disaster of 2011, there were signs of a nuclear revival in the United States and Europe, this despite the lack of permanent repositories for waste. The primary motivation was nuclear power's proven ability to reliably generate plentiful electricity without any greenhouse gases, though hardly at costs "too cheap to meter." No doubt the antinuclear backlash after Fukushima is in part temporary, but whether the revival will resume or reverse in the industrial nations is uncertain. The nuclear power construction plans for China, India, and elsewhere in the Third World will likely remain on track.

Renewables

Preindustrial societies got all their energy from renewable sources that in one form or another came from the sun. Photosynthesis converts sunshine into crops and other biomass, to be eaten by humans and animals or burned for heat. Windmills and sailboats are driven by air currents caused by uneven heating of the atmosphere by the sun. Water flows because the sun evaporates surface water, forming clouds that produce precipitation that flows via gravity from higher land to the sea.

The oil shortages of the 1970s, which made salient the limits if not the end of finite fuels, turned attention back to these energy sources that never run out. When they are used, they are not used up. This fact alone made renewables seem preferable to finite fuels. President Jimmy Carter installed solar panels on the roof of the White House. His successor and ideological adversary, Ronald Reagan, had the panels removed.

Carter and his predecessor, Richard Nixon, were the two US presidents most absorbed by the energy crisis. They saw the primary problem as America's large and growing dependence on foreign oil, especially from the OPEC nations that had recently wielded their "oil weapon" – cutting oil shipments in 1973. To reduce their vulnerability to another embargo or sudden price hike, the US and other oil-importing countries touted across-the-board efforts to become "energy independent," whether by increasing domestic production of fossil fuels, stockpiling oil in "strategic reserves," using fuels more efficiently, or pushing ahead with nuclear power. One of Carter's (short lived) initiatives was to convert coal or oil shale to "synfuel" – liquid fuel that substitutes for oil, as Germany did in World War II to make up for its shortage of petroleum. This would have produced a lot of CO_2, but in the 1970s global warming was not recognized as a problem.

The option of wholly or even largely replacing oil with unconventional renewable energy was regarded as unrealistic by many in government and industry for many reasons: they were accustomed to fossil fuels, they recognized the hugely expensive change in energy infrastructure that would be required, their businesses were dependent on fossil fuel, and the price of solar or wind energy was much higher than that of conventional fuels. There were intense debates at that time about the "hard path" (i.e., central power stations using fossil and nuclear fuels) versus the "soft path" (i.e., dispersed renewables and efficiency) to energy independence, or more pejoratively, the "hard headed" versus "soft headed" approach (e.g., Lovins 1976). California's young governor Jerry Brown, a strong advocate of renewables for his own state, was derisively nicknamed "Governor Moonbeam."

The US Congress adopted a multifaceted approach, providing encouragement for nearly every option. The critical steps for renewable energy were giving financial incentives for its developers, mitigating its high cost, and facilitating its entry into the nation's energy mix. Several nations passed legislation to subsidize development of renewables, and forced traditional electric utility companies to buy power from independent developers of renewable energy sources. This guaranteed that innovators in renewable energy would have a market for any electricity that they produced, and that they had reasonable prospects for a good profit. In the United States, however, the subsidies and other incentives diminished during the Reagan era and have subsequently fluctuated.

To date, the US has not moved very far along the renewable path, though without that initial push during the 1970s, it would not have gotten as far as it has. Several individual states have gone much further than the nation as a whole, imposing "renewable energy portfolios" that require a sizable portion of their electricity to soon be produced from renewable sources (http://www.eia.gov/cneaf/solar.renewables/page/_rea_data/table28.html).

Hydro

Flowing water is the one renewable source that has always been part of industrialization. Rushing streams turned the water wheels of England's first factories. Conversion of the kinetic energy in falling water to electricity – hydroelectricity – has been a significant part of power generation since the earliest days of the electrical industry (see Part III). Today hydroelectricity is by far the largest contributor of the renewable sources.

Niagara Falls, on the border between the United States and Canada, was one of the first large hydroelectric facilities. Upstream from the famous falls, water from the Niagara River is diverted into tunnels on both the Canadian and American sides. Out of sight, it rushes downstream through these tunnels, turning turbines that are connected to electric generators. Water that is left in the river does not produce electricity but flows awesomely over the falls and merges downstream with outflow from the tunnels. Some early enthusiasts proposed devoting Niagara's entire flow to electricity generation, eliminating the cataract; however preservationists prevailed in keeping enough water in the river to maintain the scenic view.

In most hydroelectric projects, the entire flow of a river is dammed, creating a large reservoir upriver. The level of the reservoir and the flow downstream are controlled by letting more or less water cross the dam through spillways. Some backed-up water is diverted to fall through tunnels where it turns turbines connected to generators before returning to the main downstream flow of the river. Besides producing electricity, dams and their reservoirs supply irrigation water and control seasonal flooding.

The great dams that were built in the United States and Europe during the twentieth century used nearly all sites that are geologically suitable for large-scale power production, so there is limited potential for expanding this renewable source in the industrial nations (though Canada still has much untapped potential). Most technically "good" sites that do not already have dams are now off limits for scenic or ecological reasons. In the mid-twentieth century, American environmentalists successfully opposed the US Bureau of Reclamation's proposal to dam the Colorado River inside Grand Canyon National Park but lost a similar fight to stop the flooding of beautiful Glen Canyon (McPhee 1971).

With few good hydro sites available in the industrial nations, the biggest dams are now constructed in the Third World. Itaipu Dam on the border of Brazil and Paraguay (opened in 1984) and Three Gorges Dam in China (opened in 2008) are the world's largest. Over a million people were dislocated to enable filling the reservoir behind Three Gorges Dam, and the massive engineering project is likely to have severe effects on habitat and species, also possibly increasing the risk of earthquakes.

A dam is the only renewable source of energy that is subject to a catastrophic failure, rivaling the worst accidents envisioned at nuclear power plants. More than 2,200 people died when a dam at Johnstown, Pennsylvania collapsed in 1889. In 1975 China's Banqiao Reservoir suffered a cascading failure of 62 dams, killing an estimated 171,000 people, the most deaths due to a dam failure on record (http://en.wikipedia.org/wiki/Banqiao_Dam). Nonetheless, the specter of a nuclear tragedy haunts people's imaginations far more than the prospect of a dam failure (Weart 2012).

Large dams can have serious ecosystem effects, among them preventing salmon from making their yearly return upstream to spawn, threatening their existence and the economy of fisheries. In a slight counterpoint to a century of dam building, some small old dams in the United States have been dismantled to allow fish access to their spawning grounds (Carpenter 2012). In Norway, which gets nearly all its electricity from hydro, high-voltage transmission lines that carry electricity through scenic areas to often-distant load centers are themselves objects of public opposition (e.g., Tagliabue 2010).

Wind

Little Denmark (population 5.5 million) has jumped to a leading position in wind power, getting a fifth of its electricity from the wind and hoping to raise that to 80 percent. It is home to Vestas Wind Systems, a world leader in wind technology with factories in the US and several other nations. Vestas sells its giant V90 model to "wind farms" in the American Midwest at a price of roughly $5 million apiece. The V90 stands 85 meters high, about as tall as a 40-story building, and weighs 267 US tons. At the top is a mechanical chamber containing the turbine's gearbox, generator, transformer, and control systems. It has three blades, rotating about 16 revolutions per minute and cutting a circle 90 meters in diameter. The blades spin when wind speed is over 3.5 meters per second (about eight miles per hour) and turn off when wind is too high, about 35 meters per second.

Each windmill has the capacity to produce three megawatts of power. Three hundred windmills would equal the installed capacity of one large (900 megawatt) fossil or nuclear power plant. One windmill could generate enough electricity for about 800 average American homes if it were spinning full-time, but no windmill spins all the time because the wind is not constant, sometimes too fast, sometimes too slow, sometimes not blowing at all. A wind farm that produces 40 percent of its rated output is performing well (Warburg 2012).

Underground cables run from each tower to a substation, and that is connected to the electric grid just as a fossil fuel plant would be. However, unlike the constant output of fossil fuel or nuclear plants, output from a wind farm is erratic, rising and falling with the wind. This is a problem for electric grid operators because electricity is not stored; supply must equal demand at each moment. When a strong wind stops, output drops, requiring the operator to bring replacement power to the grid or to shut off some of the users. Therefore windmills must be integrated with steadying power sources to prevent instabilities in grid operation. This is a manageable problem so long as a grid does not get most of its power from erratically changing sources, but it does require constant monitoring of the output from the wind farm and compensating adjustment as needed.

Hundreds of large windmills spread over the landscape are a remarkable sight. Cows may be grazing between the towers. Some people love the visual effect, others hate it. Offshore windmills are used by several European nations bordering the North and Baltic Seas, partly to exploit steady winds and partly to minimize visual impact. Cape Wind, a wind farm to be located nearly eight km (five miles) off Cape Cod, Massachusetts, covering 62 km^2 (24 square miles) of ocean with windmills half again as high as the V90, provoked fierce opposition because of its effect on the view from shore.

California under Governor Jerry "Moonbeam" Brown offered generous incentives to renewable energy developers. For wind development, the state picked three mountain passes where cool air blew in from the Pacific. Over 17,000 windmills of diverse size, design, and manufacture were installed, often by small, inexperienced companies rushing for a piece of the action. They did not work very well. With American windmills failing at an alarming rate, there was an opening for more reliable Danish windmills so Vestas entered the US market. But by the mid-1980s, American subsidies for renewables were evaporating. Vestas, by then heavily invested in faltering California, declared bankruptcy in 1986. The company might have vanished had it not found new markets in Europe and India. Germany, influenced by the Greens, passed a law in 1991 requiring utility companies to pay solar and wind generators favorable fees for electricity, soon making Germany the world's number one wind generator. Taking similar measures, Spain ranked second.

California's Altamont Pass, one of the state's three passes targeted for wind development, became infamous as the site where closely packed windmills killed tens of thousands of birds including the federally protected golden eagle. The destruction of birds and bats is perhaps the central environmental objection to windmills. The problem may be lessened for low-flying creatures by choosing sites for wind farms that do not interfere with migratory flyways, and by replacing small, closely spaced windmills with fewer strictures, more widely spaced, taller, and rotating more slowly. Unfortunately, taller turbines with longer blades increase the collision risk to higher flying birds, especially at night. As the number of windmills increases, the toll on flying creatures, some of them listed as endangered, will surely rise. Though windmills are hardly unique in harming wildlife, the destruction of birds and bats seems an inevitable trade-off for wind power, along with the visual presence of so many structures along rural landscapes and coastal seascapes.

Like virtually all power facilities, windmills suffer some NIMBY ("not in my backyard") opposition. For every farmer who profits from leasing land to wind developers, there are neighbors who do not want the large windmills nearby. Some find the structures and associated power lines unsightly, an industrial blight on their pastoral scenery, or are disturbed by the noise of the rotors. As with all renewable resources exploiting low-density energy, the NIMBY problem is magnified because so many collectors, spread over a large area, are needed to generate the same amount of power produced by a single fossil- or nuclear-fueled plant.

Solar

When I first visited Israel in 1979 it seemed that every roof had a solar hot-water heater, fed by the local water supply. Technologically simple, these differed little from the solar water heaters used on Los Angeles rooftops at the beginning of the twentieth century. Each had a rectangular glass-covered collector connected to an elevated hot-water tank. Water circulated through black tubes coiled under the glass, where it was warmed by the sun. Becoming lighter as it grew hotter, the water rose to the top of the collector and into the storage tank. The consumer in an apartment below depended on gravity to draw hot water from the tank as needed. Cooler (therefore heavier) water added at roof level flowed into the bottom of the collector where it was heated, and so on, circulating by natural convection. The insulated storage tank kept the water warm at night. In Israel's sunny climate, these systems sufficed for normal needs. Each tank contained an electric heater in case the sun's heat needed supplementation.

The solar water heaters on Israel's roofs were implanted in a forest of television antennas, the stubby water tanks – some garishly painted – projecting above the glass collectors. These made Israel's skyline truly ugly, even marring views of the historic walls around Jerusalem's Old City. And they were not risk free. Lacking thermostats, solar heated water could become very hot, scalding the user. Installers using rope and pulley to haul heating units to the roofs of apartment buildings sometimes let one go. Occasionally a stiff wind would catch a collector like a sail, throwing it off the roof. Despite these problems, the heaters were popular because they were inexpensive, their long-term operation far cheaper than heating water by burning fuel.

At that time Israeli engineers were already improving the aesthetics of solar heaters, installing streamlined and unobtrusive versions on larger buildings. With the replacement of TV antennas by cable, the visual obscenity was expurgated. Today, heating water is still the simplest, most cost effective "active" use of direct solar energy. Where natural gas, the major alternative, is cheap, solar is not as obviously advantageous as it was in Israel.

"Passive solar" refers to utilization of the sun's energy without specially designed collectors. It is particularly useful for reducing the need for space heating and air conditioning through proper building design, though until recently passive solar was barely considered by architects or home builders. In middle Northern latitudes, large windows should face south to catch maximum sunlight during the winter, naturally warming the interior. The roof above the windows should have large overhanging eaves. Because the winter sun stays low to the horizon even at noon, its rays are not blocked by these eaves. In summer, when the sun is high in the sky, the overhanging eaves keep the windows in shade, preventing the interior from overheating. Other passive techniques – some simple, others sophisticated – keep the interior warm in winter even after sunset, or improve natural air circulation, improve interior lighting, or make the interior space "feel" more comfortable. Passive solar adds little of anything to the cost of constructing a home or larger building and is as close as any energy source comes to being a "free lunch." Still, it is not extensively used, in part because it clashes with traditional construction practices.

Most attention today is on photovoltaics – using the sun to produce electricity. The methods may be divided into large central systems, akin to a fossil or nuclear fueled power plant, or small distributed systems, akin to Israel's rooftop solar heaters.

All of the electricity generation we have discussed so far uses a common principle: Some force – whether steam produced by burning fossil or nuclear fuel, or wind or falling water – turns a turbine that is connected to a generator. The generator itself spins a loop of wire through a magnetic field, which produces electric current in the wire.

Photovoltaic panels (arrays of photovoltaic cells), such as might be placed on the roof of a home or building, use a wholly different physical principle: the direct conversion of light into electricity at the atomic level. Some materials exhibit a property called the "photoelectric effect" that causes them to absorb photons of light and release electrons. These free electrons are collected to produce a direct current, which can easily be converted ("inverted") to alternating current.

Albert Einstein won a Nobel Prize for his explanation of the photoelectric effect. The first photovoltaic cell, built by Bell Laboratories in 1954, was mostly just a curiosity because it was too expensive for widespread use. In the 1960s, the space industry made serious use of the technology to produce electricity from sunlight aboard spacecraft, and at about the same time it found a consumer market as the light meters in cameras. With the push for renewable fuels caused by the energy crisis, photovoltaic electricity became a favored candidate. As the technology advanced, its reliability was established, and the cost began to decline, though it is still expensive compared to generating electricity by burning coal or natural gas.

In isolated sunny areas, photovoltaic cells can operate as stand-alone sources. They require only a connected battery that is charged when the sun is shining and becomes the power supply when it is dark. Once in place, solar cells operate for their lifetime, perhaps 30 years, causing no pollution, CO_2, or additional cost except maintenance. (Batteries and inverters that change DC output to AC must be changed periodically, and in desert or dusty areas, panels must occasionally be cleaned to operate at full efficiency.)

Even in deserts, sunshine is a low-density power source, i.e., the amount of power reaching a square meter of collector surface is relatively low. As a consequence it takes a very large area of photovoltaic panels to produce the amount of electricity that is generated by one conventional power plant, and of course, they produce only when the sun is shining.

In one way, cities in sunny areas are ideal platforms for distributed photovoltaics. Urban rooftops are an enormous unexploited area that could hold solar panels. As typically implemented, solar panels on the roof of a home do not power the home directly but feed their output into the electric grid to which the home is already connected. By law in many areas, utility companies are obligated to give homeowners credit for any electricity that they feed into the grid. Through a system of accounting called "net metering," the utility company measures the home's electrical input to the grid each month and subtracts this from the amount of electricity the household takes from the grid each month. The monthly utility bill charges for the net difference. In theory, homeowners who input more electricity to the grid than they use should make a profit, but in practice their credit is usually limited to the amount of electricity that they use in one year. An advantage for the homeowner of connecting to the grid is that power is available 24 hours per day, not just when the sun is shining.

The 1980s vision of small photovoltaic panels distributed over every roof has been joined if not replaced by a goal of fewer but much larger central-station generating plants located in very sunny climates such as the American Southwest or southern Europe. The primary rationales for central station solar, compared to distributed photovoltaic panels, are that it is cheaper per kilowatt of installed power, that it is easier for one company to build a central station than for a multitude of homeowners to each engage their own installer, and that the owner of a central solar plant can more easily enter into a power purchase agreement with a utility company to guarantee the sale of electrical output.

The most direct evolution from distributed to centralized solar is to assemble enormous arrays – row after row – of photovoltaic panels wired together. Though the aggregation of collectors requires much more land area than a fossil fuel plant of the same capacity, once installed the solar plant requires no fuel and produces no pollution or CO_2. Large solar power plants connect to the grid like any other power plant.

A different design for central-station solar uses parabolic mirrors (heliostats) as collectors rather than photovoltaic panels. A large circular array of mirrors focuses sunlight onto a boiler at the center of the circle, heating it to high temperature. This makes steam that, as usual, spins a turbine connected to a generator. The mirrors are computer controlled to continually adjust their position as the sun crosses the sky, maximizing the concentration of sunlight on the boiler. For a fine series of photos showing the construction of one circular array of mirrors for the Ivanpah solar project in California's Mohave Desert, see http://www.nytimes.com/interactive/2012/06/17/magazine/the-largest-solar-farm-in-the-world. The Ivanpah project, comprising three circular arrays, uses 350,000 mirrors on 15 square kilometers (5.6 square miles), which makes it the largest solar plant in the world. In full sunlight it has the capacity to generate 370 megawatts of power, equivalent to a moderate-sized fossil fuel power plant. Next door in Nevada's desert, northwest of Las Vegas, the Crescent Dunes Solar Energy Project uses a circular array of 10,000 mirrors spread over five square kilometers (two square miles) to heat molten salt in a tower rising 165 meters (540 feet); the molten salt produces steam to turn a turbine and generator, with an installed capacity of 110 megawatts in full sunlight. Heat retained in the molten salt allows this plant to continue generating electricity after sundown.

Biofuels

Unlike other energy sources described in this chapter, biofuels derived from recently grown plants (or animals that eat plants) do emit CO_2 when burned. They are carbon neutral only if new plants are grown that take out of the atmosphere the same amount of CO_2 that was put in during burning. This is well-illustrated with wood, the most traditional biofuel. Wood fiber is a hydrocarbon. Its burning (fast oxidation) or rotting (slow oxidation) produces CO_2 and H_2O along with heat. When a new tree grows, it takes carbon from the air to make new fiber, so there is no net addition over the full cycle of oxidation plus photosynthetic regrowth. If a forest is cut down and not regrown, most of the weight of the lost trees is added to the atmosphere as CO_2, which is why rainforest destruction is a potent source of greenhouse gas.

The two major kinds of biofuels are *ethanol,* a form of alcohol favored in the United States as a substitute for gasoline; and *biodiesel,* used mostly in Europe to supplement normal diesel fuel made from petroleum.

Ethanol

Ethanol is an alcohol made by fermentation of sugary or starchy crops. It is most commonly made as vehicle fuel from corn or sugarcane. Ethanol can be burned in pure form if engines are suitably modified, but often ten percent ethanol is added to gasoline so that the mix can be used in conventional engines.

The yeast-driven process of fermentation is familiar from the production of alcoholic beverages and starts with a mash of any starchy or sugary plant (barley or wheat for beer, grapes for wine, potatoes or grain for vodka, rice for sake, juniper berries for gin, sugarcane for rum, etc.). Yeasts that are added to the mash or are naturally present convert sugar into ethanol with CO_2 as a waste product. Since ethanol evaporates at a low temperature, it is easily separated from residual mash by heating. Once separated, the gaseous ethanol is cooled to return it to a liquid. (This is the process used in a bootlegger's still.) When burned, ethanol releases about two-thirds as much energy as the same volume of gasoline. When the crop is regrown, it removes from the air the CO_2 that was produced during fermentation and oxidation.

The more sugar in the mash, the more efficient the conversion to ethanol so sugarcane is an excellent feedstock. Brazil, with a tropical climate suited to growing sugarcane, turned to ethanol as a vehicle fuel after the oil shocks of the 1970s. This policy was based on the low cost of sugarcane at the time, and the country's long history with the crop. The government and manufacturers experimented with various ethanol-gasoline mixes and encouraged the sale of flexible-fuel cars that can burn pure ethanol or an ethanol-gasoline mix. In recent years, ethanol provided half of Brazil's automobile fuel, but there have also been years when Brazil had to import ethanol because of domestic shortage, and when the price of ethanol exceeded that of gasoline (http://en.wikipedia.org/wiki/History_of_ethanol_fuel_in_Brazil).

In 2005 the United States, its climate unsuitable for sugarcane but ideal for corn, became the largest producer of ethanol. One rationale for ethanol production in the US is increased energy security from reducing dependency on foreign oil imports. The US Congress subsidized corn ethanol (as it has every energy source, to varying degrees), even to the extent of placing a tariff on the importation of cheaper Brazilian ethanol in order to protect domestic producers. As a result, American farmers devoted more acreage to corn, and by 2011 ethanol accounted for ten percent of the US gasoline market by volume, ostensibly reducing by that amount the demand for petroleum. CO_2 emitted to the air in burning ethanol is retrieved from the air by the next corn crop.

Corn is not as sugary as sugarcane so its fermentation into ethanol is less efficient. Besides using arable land and irrigation water that might better be devoted to food, growing corn for ethanol requires significantly more energy from fossil fuels than growing sugarcane for ethanol. Critics argue that there is no net gain from corn ethanol, that the US subsidy is a political sop to corn-producing states and an environmental mistake. It is notoriously difficult to objectively evaluate the net gain or loss from corn ethanol; different analysts come to different conclusions. Farrell et al. (2006), evaluating six representative studies, found that all showed corn ethanol to be less petroleum-intensive than gasoline, but after taking account of the fossil fuel needed to grow the corn and convert it to ethanol, all showed that overall greenhouse emissions were similar to those of gasoline. Despite being "renewable" in the usual sense, corn ethanol seems no better for global warming than gasoline, and at least from that perspective its subsidy is poor policy.

A feedstock preferable even to sugarcane would be a plant that needed no artificial input of energy or increased arable land or irrigation, perhaps weeds or an unused waste product of agriculture such as corn stalks. Unfortunately these contain little starch or sugar for fermentation into ethanol.

The structure of most plants is built of cellulose, a linked chain of hundreds to thousands of sugar (glucose) units. The cellulose itself is impregnable against yeasts trying to ferment it, but if the long chain were broken down into smaller units of sugar, these could be fermented into ethanol, a hoped-for product called "cellulosic ethanol." Researchers are seeking enzymes that will do the job of chopping cellulose into smaller sugars. Switchgrass is a major object of study today due to its high productivity per acre, but cellulose is contained in nearly every plant and could become a plentiful source of renewable alcohol, requiring relatively small inputs to produce.

Biodiesel

Biodiesel can be made by chemical manipulation of animal fats and oils or recycled oil, but most is produced from virgin vegetable oils, usually from soybeans and rapeseed, because it is cheaper. A chemical reaction of lipids (fats and oils) with alcohol produces biodiesel plus a byproduct, glycerol. Biodiesel is a liquid, similar but not identical to diesel oil made from petroleum. It is intended for use in normal modern diesel engines, but some manufacturers do not guarantee their engines when run wholly on biodiesel, so blends are commonly sold.

Some of the same problems of bioethanol plague biodiesel. It may not be a good use of agricultural resources that would better serve to grow food. The amount of fossil fuel input to production may exceed the energy output as biodiesel (Pimentel & Patzek 2005). Depending on market conditions (and subsidies), biodiesel may be more expensive than diesel from petroleum.

Geothermal energy

If you descend deep into the earth, perhaps in a very deep mine or natural cave, it becomes warmer, the result of heat conducted outward from the center of the planet. Heat is naturally brought to shallow depths by several processes, but primarily through magma intrusions or deep circulation of water along faults and fractures in the earth's crust. The amount of heat contained within the earth is so vast an energy reservoir that geothermal power is sometimes thought of as renewable or at least inexhaustible. This is not necessarily true in practical application. Heat derived from rocks at a particular site can be disrupted by earth movements, or exhausted through circulation of colder water, or underground water itself may become depleted. The Geysers in northern California, the largest complex of geothermal power plants in the world, uses subterranean steam to turn electricity-producing turbines that generate enough power for a city the size of San Francisco. Early experience there showed that steam from local heat chambers does deplete, though it can be rejuvenated by pumping in treated sewage water. Also, the condensed steam or hot water coming up from the depths contains dissolved minerals and gases, some of them undesirable pollutants. There are questions about the stability of ground after geothermal steam or hot water is tapped or recycled, possibly triggering earthquakes. Often the most promising sites are remote, requiring new transmission lines (Seymour & Kauneckis 2012).

Finding exploitable geothermal sources is costly and chancy. In Nevada, geothermal developers look for deep fissures containing hot water over 150° C (300° F). Before any exploratory drilling can start, the site must pass inspection to ensure there is no habitat for endangered species, vulnerable water supplies, or seismic instability. If drill holes miss narrow fissures by a meter, they may be unproductive. With a successful hit, hot water is pumped or flows naturally under high pressure from the fissure to the surface. Heat from the geothermal water is transferred to a closed loop of working fluid, which turns to steam and spins a turbine connected to a generator. The pressurized geothermal water never changes to steam, but having given up its heat is pumped back into the ground to replenish the reservoir. During this recirculation, and in natural hydrothermal systems, leakage through faults and fractures near the earth's surface is common.

Steel casing is cemented around the drill holes to prevent mineral-laden water that is upward- or downward-bound from seeping into the water table. Reminiscent of objections to fracking, casings sometimes fail, possibly allowing mineralized water to pollute drinking water. Unlike fracking, geothermal wells do not pump high pressure fluid underground to create new fractures. Also unlike fracking, there is little vocal opposition to geothermal wells, perhaps because they are usually drilled on sparsely populated land, and perhaps because ranchers of the Great Basin are less prone to environmental protest.

Despite its high cost and other drawbacks, geothermal energy is an attractive option for areas with accessible underground heat chambers. Highly volcanic Iceland gets a quarter of its primary energy from geothermal, using it to generate electricity, for space and water heating in most buildings, and even to heat the streets of Reykjavik during the winter (http://www.nea.is/). Unlike solar or wind, geothermal provides electricity consistently day or night.

In temperate regions without magma intrusions, it is still feasible to use the ground for centrally heating and cooling a home or larger building, and to heat the water supply. The temperature a few meters underground is moderate and fairly constant throughout the year, roughly between 10 and 16 °C (50 and 60 °F). A working fluid piped to this depth will stabilize at, say, 14 °C. In winter, when outside air temperature is near freezing, a "ground source heat pump" (GSHP) circulates heat from the ground into a building, warming the inside air to the extent that it needs little additional warmth from a conventional space heater. In summer, the working fluid carries heat from the inside air to the earth below.

The working fluid can be continuously circulated through a closed loop between the building and the ground. Or it may be contained in vertical pipes, drilled deeper. In either case, the overall effect is to moderate temperature in the home, making it cooler in summer and warmer in winter than it would be otherwise. Regular heaters and coolers are still needed in the house, but they have less burden bringing inside temperature to the desired level.

Technically speaking, GSHPs are not *geothermal* sources because they use shallow depths of soil that are heated by sunshine rather than obtaining their heat from the earth's core. The method does not work well in very hot or cold climates, but has good potential in zones of moderate temperature, especially in new construction where pipes can be laid when the building's foundation is dug. Installation costs are higher than for conventional heating and cooling systems because they require a considerable length of underground piping, or drilling of deep wells, which can be difficult and expensive, depending on ground conditions and the availability of experienced installers. Cost would decline if the systems were widely adopted by builders. Once installed, they considerably cut operating costs (http://en.wikipedia.org/wiki/Geothermal_heat_pump).

Wholly converting to renewables

I was discussing with an electrical engineer a proposal in a 2008 issue of *Scientific American* to cover 80,000 square kilometers (about 30,000 square miles) of desert in the American Southwest with photovoltaic panels, replacing most of the nation's fossil fuel-burning generators, at least during daytime. This is feasible because the Southwest has so many sunny days and so much undeveloped open space. Electricity generated renewably there could be sent via high-voltage transmission lines to far away urban centers (Zweibel et al. 2008).

My engineer friend was aware of ways our present energy practices damage the environment: climate change, strip mining and mountaintop removal for coal mining, huge oil spills, radioactive contamination from nuclear accidents, air and water pollution, and so on. He thought there would be no environmental problems from this proposal because it was only desert, which is essentially empty, barely an environment at all. (See http://www.youtube.com/watch?v=3m5qxZm_JqM for a satirical take on the same viewpoint.)

Driving across Nevada on US 50, called "The Loneliest Road in America," I could almost agree with him as there is little to see but arid land covered by sagebrush between occasional mountain ranges. And since 85 percent of Nevada is federally owned land, why not? But the desert *is* an ecosystem, and many things live in it, some of them given legal status under the Endangered Species Act (ESA) of 1973 as endangered or threatened with extinction, usually because their habitat is disappearing.

There is currently considerable controversy in Nevada over the greater sage-grouse, a bird about two feet tall that is dependent on sagebrush for food and cover. The males are notable for bright yellow air sacks on their breasts, which they inflate during mating displays. Based on the best available scientific information about diminishing bird numbers and habitat, the US Fish and Wildlife Service considers the greater sage-grouse a candidate for ESA protection. Some people care little about the sage grouse, or about saving species' generally, but to list or not list it has become a political football. Wherever one chooses to site energy facilities, even the desert or the ocean, there are environmental issues. And there are social issues. Any major energy project can provoke controversy, with very high-voltage transmission lines nearly as pregnant targets for protest as nuclear power plants.

The next year *Scientific American* ran an even more ambitious article: "A Plan to Power 100 Percent of the Planet with Renewables" (Jacobson & Delucchi 2009). Within two decades, the authors calculated, all human power needs could be met using only wind, water, geothermal, and solar energy, with no contribution from fossil fuels, biofuels, or nuclear power. There are other such plans. They typically begin by noting the truly enormous amount of energy that reaches the earth every day as sunshine, pointing out that the total amount of energy generated daily by humans for our own use is a tiny fraction of that amount, and concluding that with technologies now (or nearly) available, we could in the near term harvest that tiny fraction from sunshine in its various manifestations. We may presume that these are purely intellectual exercises rather than serious proposals because it is hard to conceive of anyone but a fanatic wishing the world to undertake so costly and risky an experiment that might not be successful and would be difficult to undo.

Such exercises are useful in expanding our thinking and questioning assumptions. For example, in what ways would the world be better off if its energy sources were completely renewable, as opposed to being wholly dependent on some other source (e.g., coal with carbon capture, nuclear power, a hydrogen economy), or more realistically, on a diverse mix of energy sources? If one's sole concern were depletion of finite fuels or global warming, then carbon-free renewables (or nuclear power) would be admirable options. If one were more concerned with monetary cost; or with protecting open land, free rivers, and offshore seascapes from developers of energy farms; or with ensuring reliable supplies of energy throughout the day and year, then being wholly dependent on renewables might be disadvantageous.

Hundreds of readers of *Scientific American* posted diverse technical objections to the plan, as they had to the prior year's article. They rejoined that authors had ignored environmental issues, finessed the intermittent nature of renewables by assuming that they could be smoothed out (using solar during the day, wind at night), were unrealistically optimistic about the ease and cost of implementation, barely addressed the immense network of transmission lines needed to transfer electricity from renewable sources to end users, ignored the enormous amounts of material and energy required to manufacture and install all the collectors of energy from low-density sources, and so on. As a sociologist, I wondered how many people and governments would favor the spending to totally convert to renewables, how much social disruption would there be in doing it so quickly, how many parties would protest one or another aspect of the implementation, and why would any energy planner want to do it anyway? Stipulating that we burn too much coal and oil, does it follow that we should not burn any? Would global warming be discernibly worse if we continued to use a modest amount of natural gas?

Conclusion

Compared to the finiteness of fossil fuels, all of the non-carbon energy sources in this chapter offer long-term if not eternal availability. Using breeder technology, a nuclear power reactor creates new plutonium fuel as it fissions uranium fuel. Geothermal heat from the earth's core is limitless in supply. Renewable sources based on sunlight – wind, hydro, solar, biomass – are never used up. If fuel depletion and global warming were our only problems, non-carbon sources would be top choices.

But these are not the only considerations. Every non-carbon source has disadvantages. The point need hardly be made for nuclear power, but the negatives of solar, wind, or hydro are often overlooked or discounted. Failure of a nuclear reactor may seem more ominous than collapse of a dam. In fact, dam failures have killed far more people than nuclear reactor accidents. Intact dams block rivers and create reservoirs with their attendant problems. Solar and wind are sometimes unavailable or intermittent, depending on the latitude and climate at a site as well as the time of day and season of the year. Geothermal is exploitable only where magma intrudes near the surface. Because many of the best sites for renewable or geothermal energy are distant from load centers, their extensive use implies an enormous expansion of electrical grids, which themselves cause environmental insult and public complaint. And energy from non-carbon sources usually costs more money than energy from fossil fuels. Switching entirely to non-carbon sources would not free us of energy problems.

7

WAS GROWTH OF ENERGY AND ELECTRICITY USAGE IN INDUSTRIAL NATIONS DUE MORE TO POPULATION GROWTH OR TO OTHER CAUSES?

Looking across the world's nations, the principle determinants of energy consumption are population size and affluence (York et al. 2003; Ward 2006; Rosa et al. 2009). In the imagery of *New York Times* columnist Thomas Friedman, the world's resource problem is not due simply to too many people, but to too many people living a high-consumption "American" lifestyle (2008). The relative importance of these drivers has been a perennial point of contention, with neo-Malthusians giving most weight to population growth while others like Friedman emphasize our culture's extravagant use of energy.

Reviewing the past three decades of theoretical argument for and against population as the central driver of anthropocentric environmental change (of which fuels consumption is a major component), Dietz et al. (2010) note the lingering impasse. To a considerable extent the opposing positions are at cross purposes. The role of affluence, so central in explaining consumption differences between developed and less developed nations, is far less important when inquiry is limited to industrial nations that are all affluent. Also, population growth is relatively low in the developed nations. Thus, the importance of any single driver depends on whether we consider the world overall, or only the richest nations. Even within the industrialized world, the situation of one country may differ markedly from that of another.

Affluence includes both the "pull" of consumer demand and the persistent "push" of fuel providers and electrical utility companies to increase the supply and consumption of energy, thereby increasing their profits. These are not the only factors besides population that affects energy and electricity use. Other drivers are the efficiency of a nation's energy technology; its mix of fuels; its mix of manufacturing, service and transportation activities; climate and geography; its environmental regulation of energy production; degree of electrification; availability of indigenous fuels; import and export of fuels and electricity; and consumption of energy by the nation's energy- and electricity-producing industries themselves. I call these, collectively, the "nonpopulation causes."

In the second half of the twentieth century, was rising use of energy and electricity in industrial nations due more to population growth – the emphasis of the neo-Malthusians — or to these other causes?

Methods

I initially consider the 21 OECD nations larger than two million in current population (see Table 3.1). Subsequent analysis focuses on energy and electricity trends in eight of the largest industrial countries: the United States, Canada, United Kingdom, France, Spain, Italy, Japan, and Australia. These are diverse in size, location, climate, language, and culture, hence providing a good overview of industrial democracy. A conspicuous omission is Germany, where reunification in 1990 was so great a structural change that comparison of its energy trends to the other cases makes little sense.

Energy and other national data used here are from "Energy Balances of OECD Countries (2009 Edition; http://wds.iea.org/wds/ReportFolders/ReportFolders.aspx), which provides the longest time series (since 1960) of energy trends that is reasonably comparable across nations. A nation's *total primary energy supply* (TPES) comprises all energy sources input to the society including fossil fuels, nuclear power, hydroelectricity, and other renewables. *Electricity consumption* refers to the amount of electricity distributed from domestic power plants to end users + imports of electricity – exports – distribution losses.

The total energy consumed by a nation in a year (E) equals the product of per capita energy consumption (e) and total population (P). If e and P are independent of one another, then differentiating this identity produces $dE = e\ dP + P\ de$. The term $e\ dP$ is the portion of change in energy consumption that is due to changing population. Thus, the change in total energy consumption is decomposed into a population component ($e\ dP$) and a nonpopulation component ($P\ de$).

This partition is logically derived from a mathematical identity and does not necessarily have substantive meaning. Some increases in energy consumption are clearly the result of growing population, including construction of new housing subdivisions or apartment blocks, greater number of cars on the road and amount of gasoline purchased, addition of new shopping malls, and increase in energy consuming jobs. On the other hand, there is no clear rationale for distributing the enormous electricity consumption of a large aluminum smelter across the entire population when most people have no direct connection to it. For the partition to be meaningful, we must assume that each nation at a given time has a particular culture (or norm) of energy use, and that the modal person added to that nation will more or less use its per capita energy consumption (e). This is implicit in Thomas Friedman's complaint that too many people live an "American" lifestyle. I proceed on this assumption, recognizing that it will be justified or not depending on its utility and the evaluation of other researchers.

All terms in the equation are estimated from historical statistics. The differential equation is treated as a difference equation with the year as a time unit. IEA data include each nation's yearly population, energy consumption, and electricity consumption. From these, year-to-year changes in the components of the equation are calculated arithmetically (Mazur 1994).

Results

Figure 7.1 compares increases in population, in TPES per capita, and in electricity consumption per capita from 1960 to 2008 for 21 OECD nations. Mean increases were 142 percent for population, 320 percent for TPES per capita, and 677 percent for electricity consumption per capita. With only three exceptions (US, Australia, UK), per capita energy consumption increased more than population, especially in the Mediterranean region. Electricity consumption grew faster than TPES in all nations except Norway, and again the increases were greatest in southern Europe.

We will look more closely at the eight countries with current populations over 20 million (excluding reunified Germany), which together account for over 80 percent of the TPES and electricity used by the industrialized world. Figures 3.5 and 3.6 show trends for these eight in per capita consumption of TPES and of electricity, respectively. Canada and the United States are by far the highest per

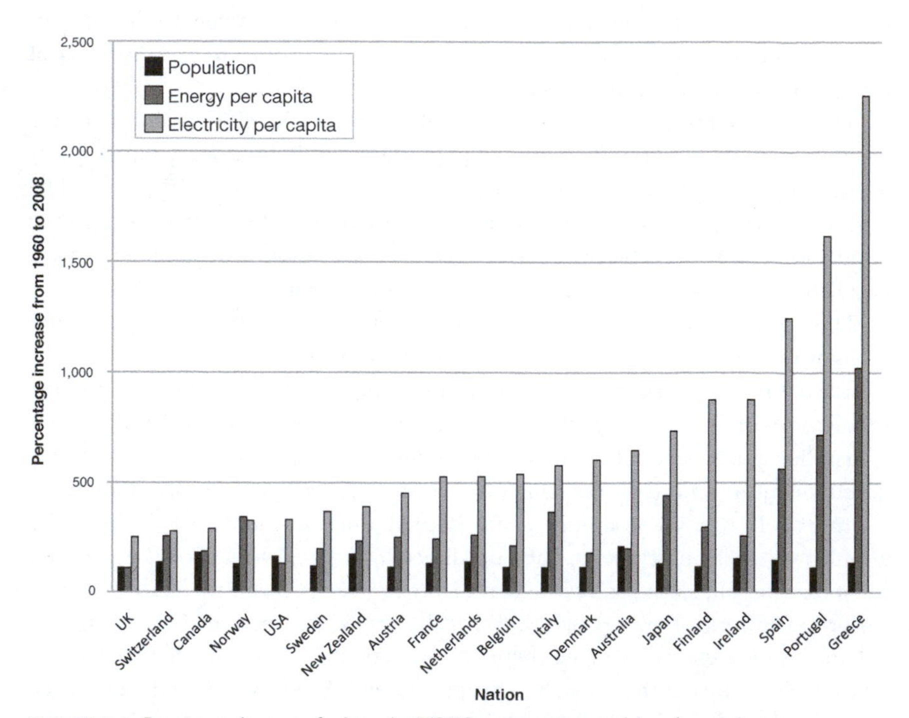

FIGURE 7.1 Increases in population, in TPES per capita, and in electricity consumption per capita from 1960 to 2008 for 21 OECD nations

capita users of primary energy, and Canada outstrips even the US in per capita use of electricity. These two nations have similar cultures and an interconnected energy system, so the similarity of their trends is expected. Australia occupies a middle position. The remaining nations – the Europeans and Japan – are clustered at the lower end of the range (with Spain and Italy lowest).

Population and nonpopulation components were calculated for each year's change in TPES. Components were normalized by dividing them by TPES that year, thus eliminating their connection to the nation's quantity of energy used (i.e., for each year, the normalized population component = *e dP/TPES*; the normalized nonpopulation component = *P de/TPES*). These components for the US are plotted by year in Figure 7.2, which in broad view is similar to component plots for the other nations (not shown).

The population component graph is smooth and nearly level, reflecting the fairly continuous year-by-year increase in US population. In contrast, the graph of the nonpopulation component is erratic with the absolute magnitude of yearly fluctuations averaging twice the size of fluctuations due to population. For all eight nations, the mean absolute value of nonpopulation components is 3.5 times the mean population component; in most years it ranges from two to six times the size of the population component. Thus, overall, population was less important than nonpopulation factors in driving TPES upward in most years, however there were important changes by decade. During the 1960s, in the US and elsewhere, nonpopulation factors produced yearly increases in energy consumption.

FIGURE 7.2 Yearly changes in population and nonpopulation components of US change in TPES

In recent years, nonpopulation causes are nearly as likely to diminish TPES as to increase it. The population component, though typically smaller than the nonpopulation component, invariably moves energy consumption upward. Perhaps population growth, integrated over many years, dominates nonpopulation factors – an example of "slow and steady wins the race." To evaluate this possibility, yearly changes in TPES (normalized) due to each component (population and non-population) were summed over the years since 1960 (with 1960 set as a zero point). These cumulative changes for the US are shown in Figure 7.3. Because population growth is fairly constant, its cumulative contribution moves smoothly upward.

Nonpopulation factors in the US have a more interesting history, which is seen very clearly in Figure 7.3. Energy consumption (net of population) rose rapidly during the 1960s and early 1970s until interrupted by the OPEC oil embargo of 1973. After resuming its upward trend by the mid-1970s, there was a far greater downturn after the Iranian revolution of 1979. This initiated a period of modest growth (net of population) until the end of the century. Since 2000, while energy consumption due to population continues to rise, growth from nonpopulation factors actually declined. That is, apart from population growth, the United States has finally begun to reduce its energy consumption.

Similar plots for the other seven nations are shown in Figure 7.4, a through g (with vertical axes again normalized on TPES, beginning at zero in 1960). In Canada

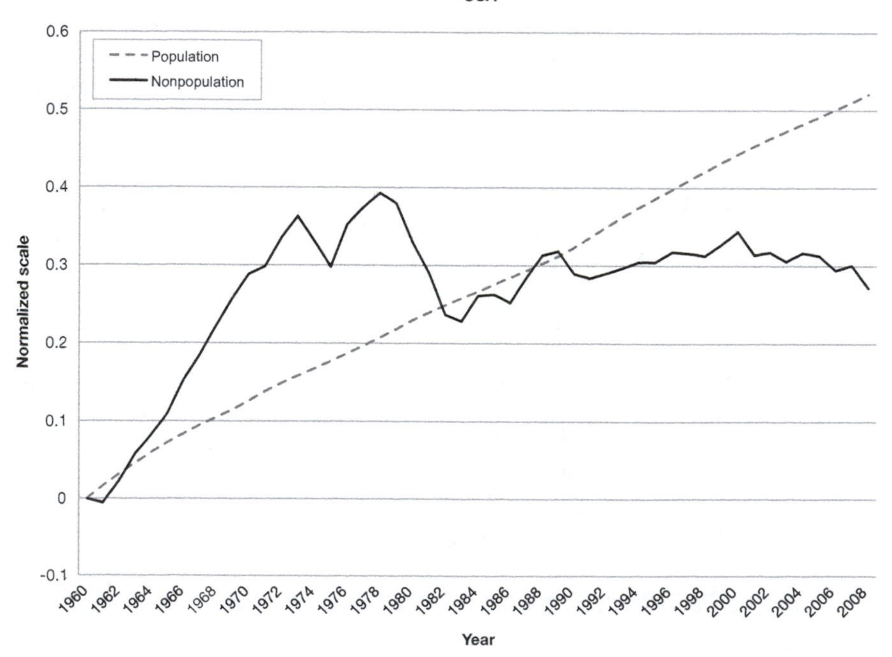

FIGURE 7.3 Cumulative changes in US TPES (since 1960) due to population and nonpopulation factors

and Britain, unlike the US, nonpopulation factors dominated population contributions to TPES until the last few years. In Australia, where the rate of population increase is most rapid, population is equally as important as nonpopulation factors in driving TPES upward. The European nations (France, Spain, Italy) and Japan show similar pictures, with nonpopulation factors far more important than population as a driver.

The nonpopulation graphs in Figures 7.3 and in 7.4 a–g, are especially useful in showing how differently external perturbations affected these different countries – effects that are not so clear in the trends for TPES per capita. The oil embargo of 1973 was aimed primarily at nations that supported Israel in the Yom Kippur War of that year, including the US, Canada, the UK, and Japan (Licklider 1988). These impacts are easily seen in Figures 7.3 and 7.4. The embargo had little or no immediate impact on energy consumption in Spain, Italy, and Australia. The United Kingdom was particularly crippled in 1973, because the embargo combined with strikes by British railroad workers and coal miners, contributing to a change in government, and again by the Iranian revolution of 1979. The post-2000 downturn in energy consumption (net of population growth), coinciding with the Iraq war and rising energy prices, was more pronounced in the UK than in the United States, while Australia is the one nation with increase in consumption unabated.

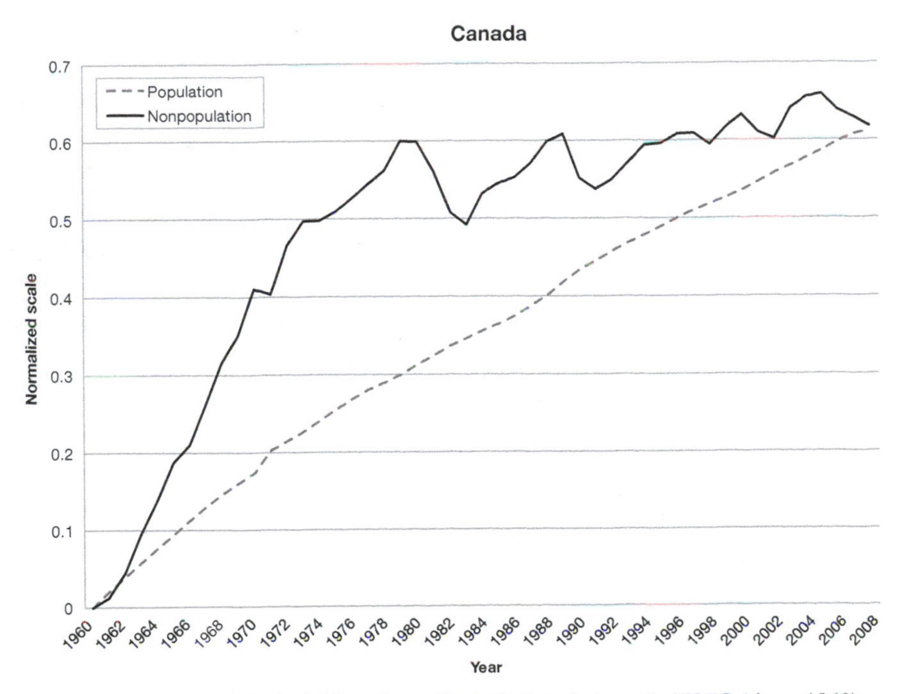

FIGURE 7.4 A–G Seven industrial nations: Cumulative changes in TPES (since 1960) due to population and nonpopulation factors

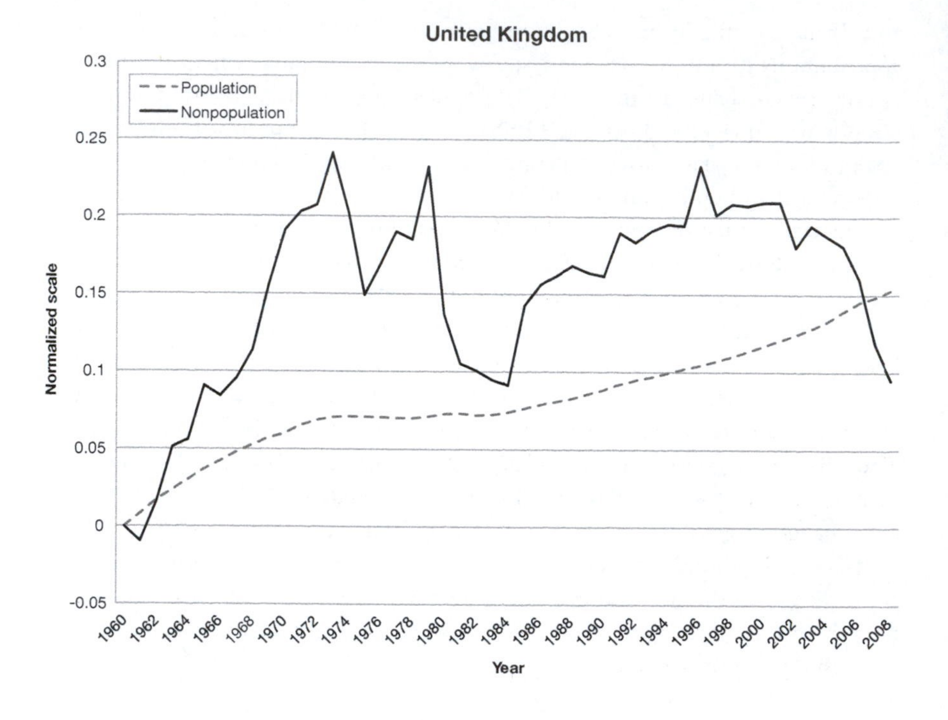
United Kingdom
Population
Nonpopulation
Normalized scale
0.3
0.25
0.2
0.15
0.1
0.05
0
-0.05
1960
1962
1964
1966
1968
1970
1972
1974
1976
1978
1980
1982
1984
1986
1988
1990
1992
1994
1996
1998
2000
2002
2004
2006
2008
Year

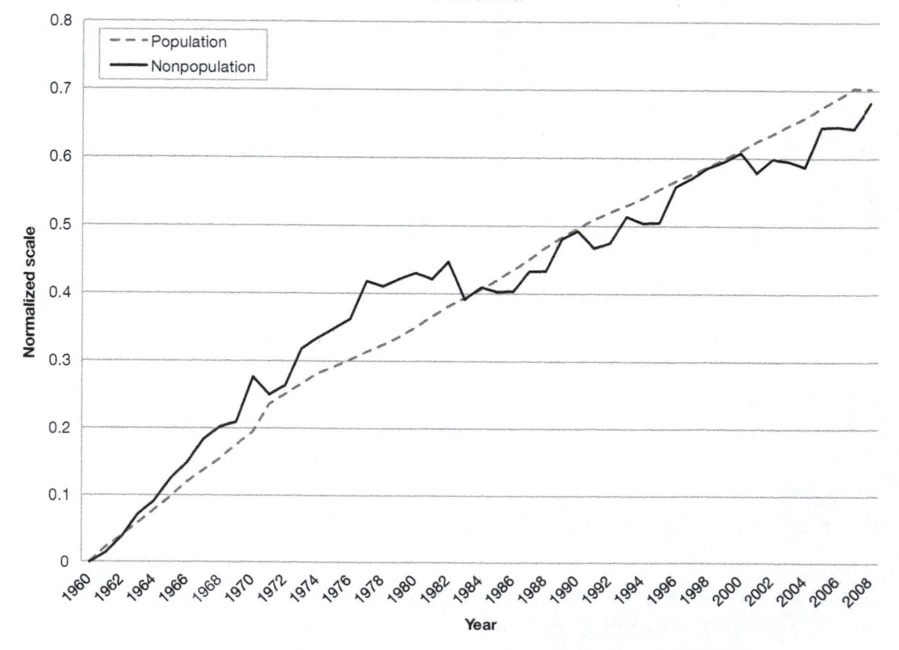
Australia
Population
Nonpopulation
Normalized scale
0.8
0.7
0.6
0.5
0.4
0.3
0.2
0.1
0
1960
1962
1964
1966
1968
1970
1972
1974
1976
1978
1980
1982
1984
1986
1988
1990
1992
1994
1996
1998
2000
2002
2004
2006
2008
Year

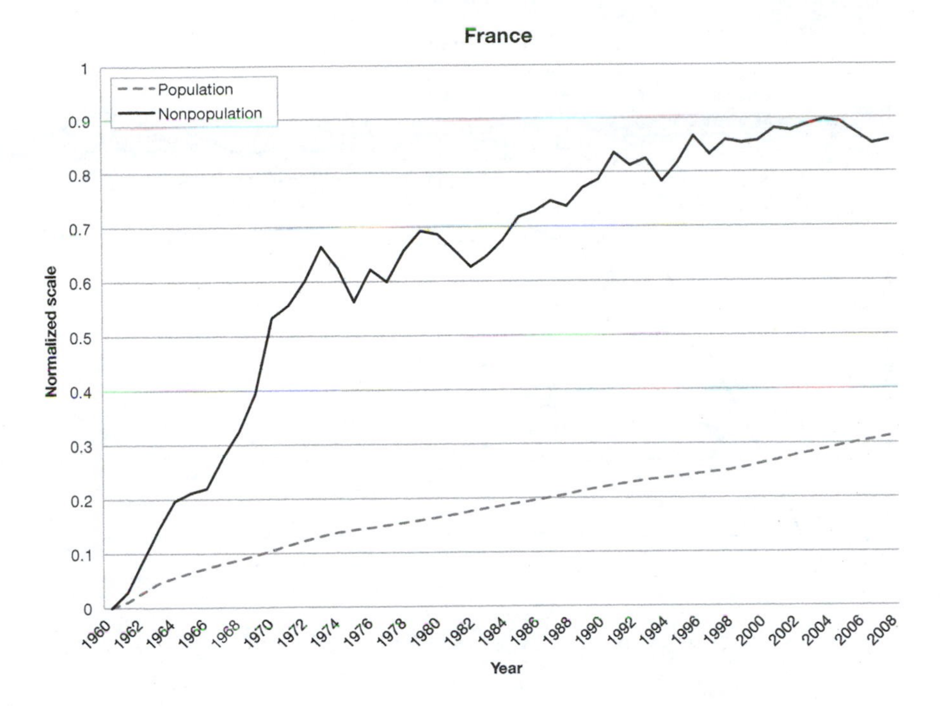
France
Population
Nonpopulation
Normalized scale
0
0.1
0.2
0.3
0.4
0.5
0.6
0.7
0.8
0.9
1
1960
1962
1964
1966
1968
1970
1972
1974
1976
1978
1980
1982
1984
1986
1988
1990
1992
1994
1996
1998
2000
2002
2004
2006
2008
Year

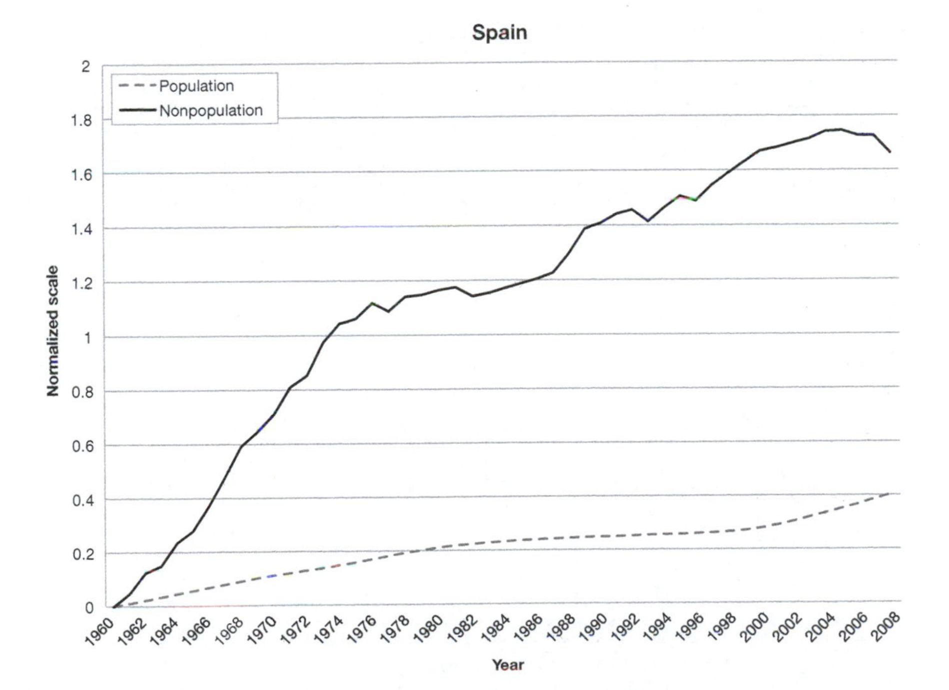
Spain
Population
Nonpopulation
Normalized scale
0
0.2
0.4
0.6
0.8
1
1.2
1.4
1.6
1.8
2
1960
1962
1964
1966
1968
1970
1972
1974
1976
1978
1980
1982
1984
1986
1988
1990
1992
1994
1996
1998
2000
2002
2004
2006
2008
Year

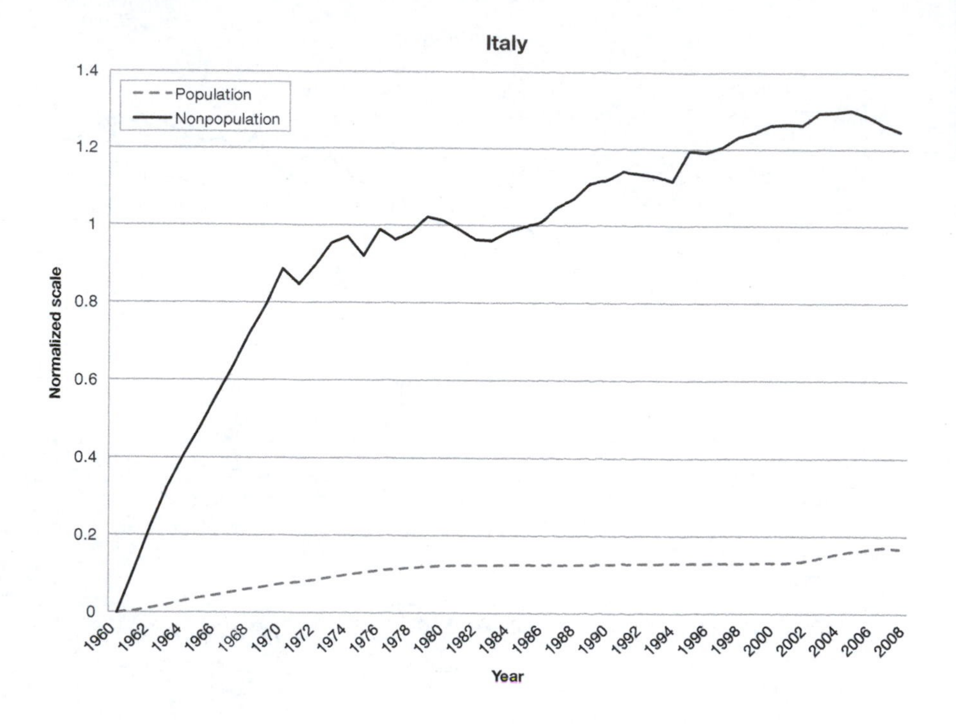
Italy
Population
Nonpopulation
Normalized scale
1.4
1.2
1
0.8
0.6
0.4
0.2
0
1960
1962
1964
1966
1968
1970
1972
1974
1976
1978
1980
1982
1984
1986
1988
1990
1992
1994
1996
1998
2000
2002
2004
2006
2008
Year

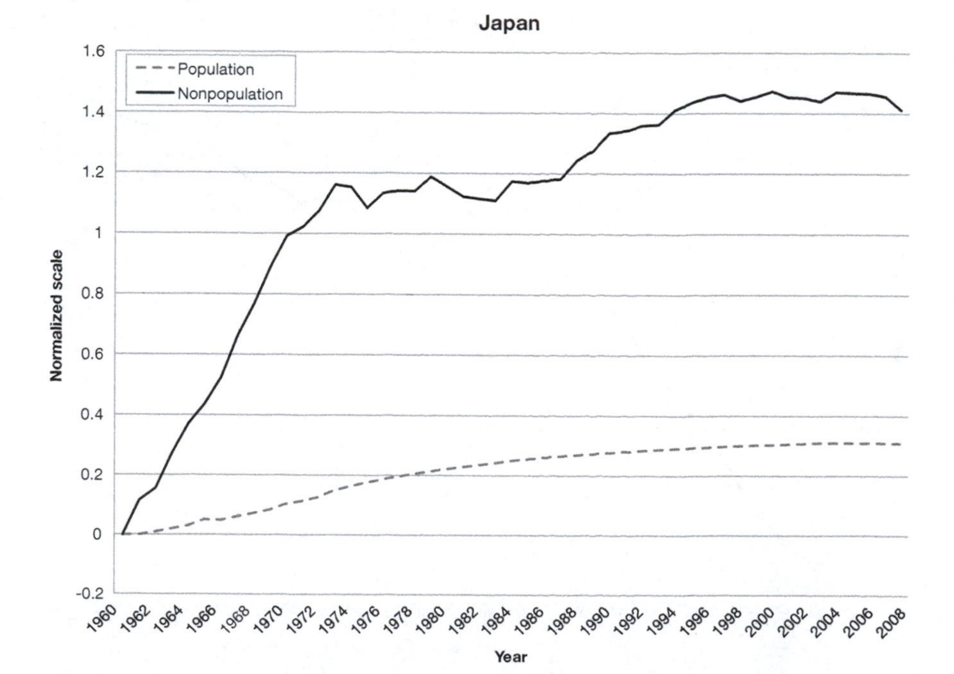
Japan
Population
Nonpopulation
Normalized scale
1.6
1.4
1.2
1
0.8
0.6
0.4
0.2
0
-0.2
1960
1962
1964
1966
1968
1970
1972
1974
1976
1978
1980
1982
1984
1986
1988
1990
1992
1994
1996
1998
2000
2002
2004
2006
2008
Year

Population and nonpopulation components of changing electricity

Growth in electricity consumption was decomposed into population and nonpopulation components by the same method. Figure 7.5 shows cumulative increases since 1960 due to both components in the US. Since the "energy crisis" of the 1970s, the nonpopulation component of US electricity growth has leveled off. Pictures for other nations (not shown) are similar. In each country, nonpopulation factors greatly dominated population in spurring growth. This is a reflection of electricity growing far faster than population (and TPES) in all OECD nations (Figure 7.1).

Figure 7.6 compares graphs of the cumulative nonpopulation component of electricity consumption for the eight nations. Electricity growth due to nonpopulation factors accelerated everywhere during the 1960s, then decelerated everywhere, and finally – in most nations – turned down slightly (as plotted on a normalized scale, though in absolute units, electricity consumption continues to rise). Short-term decreases due to the oil shocks, so salient in TPES graphs (net of population growth), are less visible for electricity, which switched to energy sources other than oil (Yergin 2011).

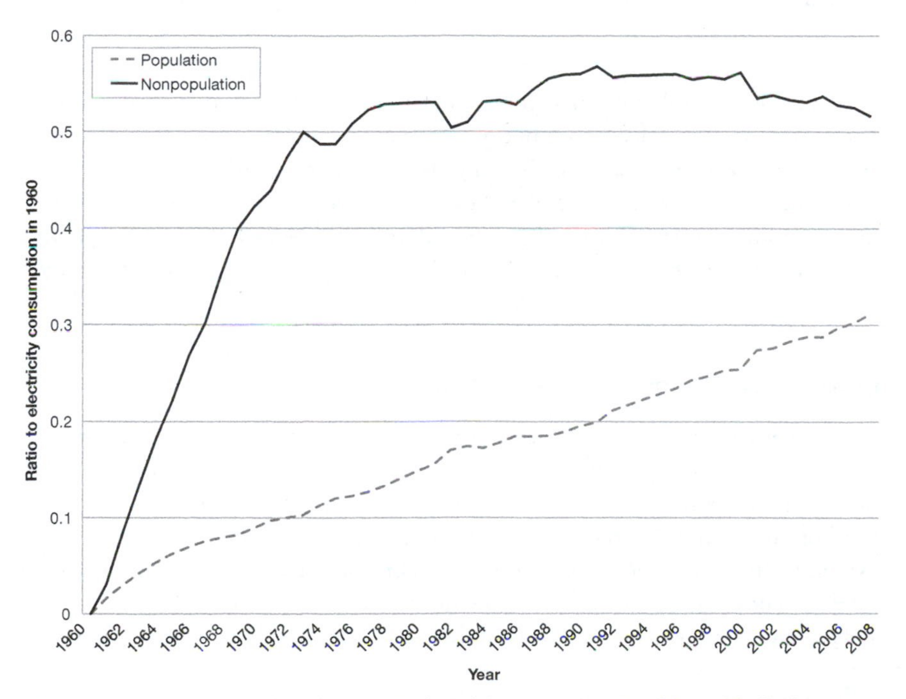

FIGURE 7.5 Cumulative changes in US electricity consumption (since 1960) due to population and nonpopulation factors

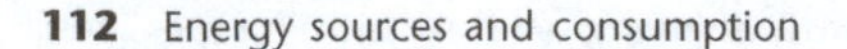

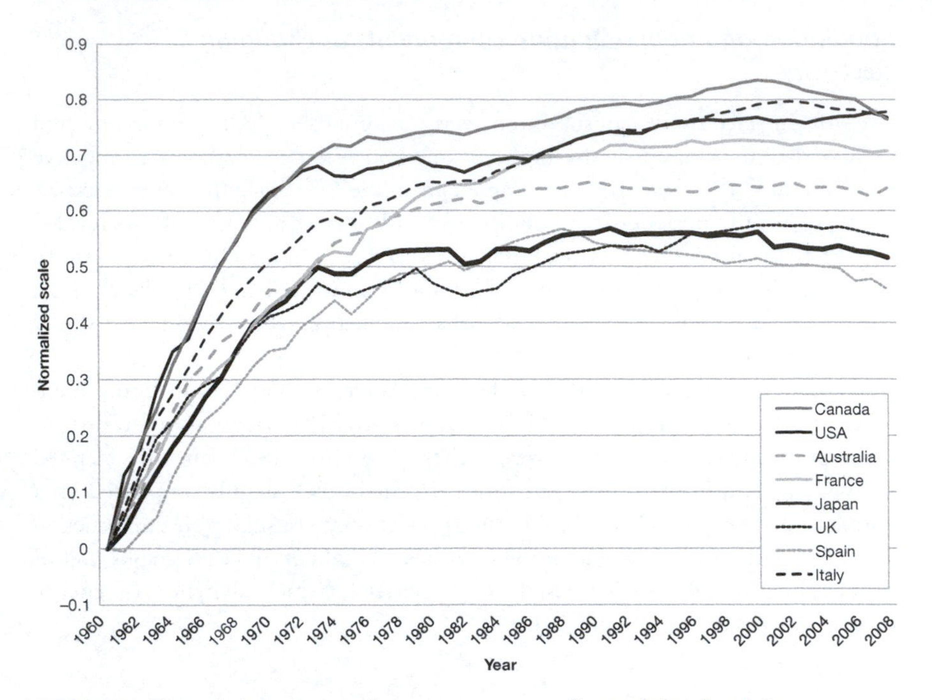

FIGURE 7.6 Cumulative nonpopulation component of growth in electricity consumption (since 1960) for all eight nations

Since the vertical scale in Figure 7.6 is normalized, thus removing absolute differences in electricity consumption among the eight nations, their relative ranking is unusual. Nations showing the highest (relative) trajectories of electrification growth since 1960 are Spain, Italy, Japan, and France – either slow starters in industrialization or badly damaged in World War II. The lowest growth trajectories are in the UK, US, and Canada, all early industrializers that emerged from the war without severe damage.

Industry and transportation sectors

In national energy accounts, diverse end users are sorted into major sectors, the most important being industry and transportation. National consumption per capita for transportation (mostly from oil) rose in all nations from 1960 to 2008. Nations that consumed the most (or least) in 1960 hold similar ranks in 2008, exhibiting the same momentum noted earlier. Americans and Canadians, with parallel trends, consistently consume the most; Australians are middling; Europeans and Japanese consume less. Impacts of the oil shocks were deepest in the US and Canada.

The industrial sector is different. Industrial consumption per capita rose in all nations except Australia until the 1970s, and then declined in most nations, most of them retaining their rank order. The American trend is a striking exception, dropping from highest per capita industrial consumer in the 1960s to a middle rank by the 1990s. Even Canada, which usually parallels US trends, maintained its momentum as a high consumer of industrial energy while its southern neighbor headed downward.

The anomalous decline in US industrial energy consumption is not easily explained. Certainly it is not part of a broad American strategy of energy conservation because similar declines did not occur in the transportation, residential, or commercial sectors. Nor is it a result of declining industrial production, which, measured by the IEA's Industrial Production Index, has been rising in the US since 1960, more or less on a par with the other industrial nations.

The US has halved its energy intensity (i.e., the amount of energy consumed per constant dollar of GDP) since the early 1970s, which is better than most of the eight nations but not as good as the UK; and it has improved its electricity intensity over the same period by 20 percent, which again is better than most of the industrial nations but about the same as Canada and less than the UK. However, to explain declining energy consumption in American industry as a result of improved energy intensity begs the question: Why has energy intensity improved? Is it due to improving technological efficiency, or to the changing structure of American industry, or to moving energy-intensive production to foreign facilities? EIA analysts conclude that aggregate improvements in manufacturing energy intensity are due to changes in the mix of goods produced. If the mix of goods is held constant, there is little if any improvement in energy intensity (Schipper 2006). The effect of moving energy-intensive industry offshore was not evaluated, but it likely has some effect on aggregate industrial energy intensity.

Conclusion

Did population growth or other causes most contribute to the rising use of energy and electricity in the industrial nations since 1960? The answer depends on the nation and how one interprets the question. Yearly fluctuations in TPES due to population growth averaged only one-sixth to one-half the magnitude of fluctuations due to nonpopulation factors. Thus, for a typical year in most nations, the population component was relatively small. However, in the 1970s and afterward, the yearly change in TPES due to nonpopulation factors was often negative, while the yearly change due to population growth, though smaller, was invariably positive. Therefore the cumulative effect of population growth from 1960 to 2008 eventually equaled or exceeded the cumulative effect of nonpopulation factors in the US, Canada, the UK, and Australia. Population growth, even viewed cumulatively, was relatively unimportant for the growth of TPES in France, Spain, Italy, and Japan. To some extent, this split is explained by relatively high population growth in Australia, Canada, and the US (but not the UK), and relatively low population growth in Japan, France, and Italy (but not Spain).

The situation of electricity is different. Here we can say without qualification that population growth was a minor driver of the rapid increase in electrification of all industrial nations, especially in the years prior to the energy crisis of the 1970s. On average, from 1960 to 2008, per capita consumption of electricity increased twice as fast as per capita TPES and nearly five times as fast as population, though rapid escalation has abated since the price rises of the 1970s.

It may seem puzzling that US population increase was an important driver of total primary energy consumption since 1960, but that population had only a modest effect on the growth of electricity consumption since 1960. This ostensible conundrum is explained by the year-to-year variation in TPES (net of population), sometimes rising, sometimes falling due to nonpopulation factors like oil prices. As a result, America's per capita consumption of TPES was lower in 2008 than in 1975. A continually larger share of America's yearly TPES was being used to generate electricity. Very soon the United States will devote half of its primary energy sources to the production of electricity, even more if there is a large move to electric cars.

A quick glance at Figure 7.3 might convey the impression that the US was uniquely frugal with energy after 1975, cutting its nonpopulation component of TPES growth more than any other industrial nation. This is correct in relative terms, but one must remember that American frugality began from an extraordinarily high base level of wastefulness, so it was relatively easy to eliminate the worst overuse by requiring, for example, higher miles-per-gallon automobile fleets. Most of the improvement was in the US industrial sector, partly due to a changing mix of goods produced, possibly also because energy-intensive industrial activities were moved to the Third World, and likely because inefficient pre-World War II industrial plants, which remained undamaged by the war and functional into the 1970s, could be closed down without serious economic loss.

New electricity-dependent technologies are an obvious reason for some of the growth in the nonpopulation component of electrification, but at the same time, other newly introduced technologies including compact-fluorescent light bulbs and kitchen appliances are more electricity efficient. Some of the increase in electricity consumption is obviously gratuitous and wasteful, like needlessly extreme air conditioning or urban lighting to the extent that now it is often called "light pollution."

PART III

Electric power

8

POWER GRIDS

Any region's electric power grid is an assemblage of high-voltage transmission lines and lower-voltage distribution lines that connect diverse generating plants (using fossil fuels, nuclear power or hydro and other renewable sources) to numerous consumers of electricity (including residential, commercial, industrial, and transportation users). All elements within each regional grid are synchronized to their own alternating current (AC), which in the United States is 60 hertz (Hz, i.e., cycles per second), but are not synchronized with other regional grids.

Electricity cannot be stored efficiently, so energy generated at any moment (and put into the grid) must match energy consumed (taken out of the grid). Since consumption varies by time of day and by season, the amount of electricity flowing through the grid is constantly monitored and adjusted. (Should there be a small mismatch between input and output, the voltage and frequency of the grid deviate from their design levels; if the mismatch increases, the grid become inoperable.)

To minimize the cost of electricity, utilities usually supply their "base load" by continually running (except for maintenance outages) their cheapest operating sources like coal, hydro, or nuclear power plants. As demand increases, more electricity is added from generators that are rapidly brought on line, especially gas turbines. Another option is to purchase excess electricity from another system, which can only occur (with minor exceptions) if the selling and purchasing systems are synchronously connected by transmission lines. Transferring power through one system for use by another system is called "wheeling."

Electrification of the United States

The British scientist Michael Faraday discovered in 1831 that an electric current is produced in a copper wire loop if it is moved between the poles of a magnet (or conversely, if the magnet is moved near the wire loop). This is the principle of an electric generator, and though it was one of Faraday's more important (of many) scientific discoveries, it brought him no profit because there was at the time no worthwhile use for electric current.

Thomas Edison recognized that money could be made if electricity powered light bulbs. It had already been noted that electric current passing through a wire filament made it warmer, eventually to a degree that the wire glowed, but very soon afterward it melted. The need, realized by many contemporary inventors, was for a glowing filament that did not fail so quickly. Edison came up with a practical solution by 1880, as did the less celebrated Briton Joseph Swan a bit earlier. There were two keys. First, to prevent oxidation of the filament, it was enclosed in a glass bulb containing either a vacuum or an inert gas, thus keeping oxygen from reacting with the wire. The second key was a matter of sheer persistence, of experimenting with hundreds of filaments of different composition until finding one sufficiently long lasting to be commercially useful.

Edison truly surpassed his fellow inventors by incorporating his light bulbs into a commercially successful system. He opened the first central electric plant in 1882 on Pearl Street in Manhattan's financial district, lighting the homes of wealthy customers and nearby businesses. He used direct current (DC) circuitry, like all previous forms of electrical technology including the telegraph, telephone, and arc lamp. The innovation spread quickly. By 1890 there were one or more electric companies in several American cities and in Europe, their primary functions to provide lighting and to power electric trolleys (Hughes 1983). Electricity came to smaller places too. The California ghost town of Bodie contains a substation that by 1892 was distributing current from a hydroelectric plant in the mountains, 20 kilometers (13 miles) distant, to the town's major mining operation.

Almost from the outset, the "battle of the currents" pitted Edison's DC against alternating current championed in the US by George Westinghouse (McNichol 2006). Edison, nearly as adept a propagandist as an inventor, tried to portray AC as the more dangerous current. At the time, the New York State legislature was seeking a more humane way than hanging to execute criminals, and at Edison's urging moved to try electrocution by alternating current. An inventor named Harold Brown, working in cooperation with Edison, acquired a Westinghouse generator for New York's Auburn Prison and in August 1880 watched the first person legally executed, receiving two prolonged jolts of AC before he died. Westinghouse commented that the man might better have been killed with an axe. Edison's publicists asked if people wanted the executioner's current running through their streets into their homes. (Actually it is the amount of current passing through the body, rather than its alternating or direct nature, that causes death.)

Edison and Brown killed a series of ever larger animal via AC, a fate Edison called being "Westinghoused." In 1903 a circus elephant named Topsy became available after squashing three handlers and being deemed too dangerous to keep alive. On her day of execution, after Topsy was fed cyanide-laced carrots to ensure her demise and in front of an estimated 1,500 spectators, she received a high dose of AC that brought her down, though it is uncertain if it caused instant death.

Edison's movie of the event – he was an inventor of the motion picture camera – was released the next year and seen in the United States and Europe. The film is widely available on the web, e.g., http://www.youtube.com/watch?v=ZCx89BRbVeU .

Propaganda aside, the crucial issue came down to cost of transmitting electricity over long distances. DC traveled through copper lines at a fixed voltage (110 volts) and relatively high current (amps), causing unwanted heating of the wires, wasting energy and money (Platt 1991). With AC, a transformer near the generator could raise the voltage to a higher value for transmission over long distances, enabling the same amount of electricity to be carried at lower current, hence less heating of wires and less waste of energy. Near the point of use, another transformer at a substation lowered the voltage before distribution to customers. When the first power station at Niagara Falls began generating in 1895, it supplied AC to electrochemical factories sited along the river gorge, and the next year began sending electricity to Buffalo, New York, 32 kilometers (20 miles) away. AC soon became the nearly universal mode. In 1910, on the Canadian side of the falls, power was sent 130 kilometers to Toronto via a government-owned transmission line. Soon Niagara's hydroelectricity was traveling 650 kilometers (400 miles) to New York City.

Growing to maturity: 1910–60

Samuel Insull (1859–1938) is accorded primacy in shaping the organizational form that dominated America's electric power industry during most of the twentieth century: the investor-owned utility (IOU) regulated by an agency of the state in which it operated. Born in London, he moved to the United States as a young man and became Edison's personal secretary, helped set up the Pearl Street station, and soon was vice president of the Edison-derived General Electric Company. In 1892 the 32-year-old Insull moved to Chicago, then the major and fastest growing city in the Midwest, as president of the Chicago Edison Company, remaining its head for four decades. He was a wheeler-dealer who within 15 years bought out the city's 20 other electric companies with their franchise service areas, often junking their inefficient equipment and tying them together into a monopolistic, mass-producing, technologically efficient, and economically operated network covering the entire metropolitan area. Later he acquired outlying suburban companies and those in neighboring cities.

Electricity provided by a central station was the mode preferred by Insull and Edison. In the early years, this had to compete with gaslight, kerosene, electric arc-lamp systems, electric generators owned by trolley companies (the largest consumers of electricity), and small self-contained generating systems. Insull reduced the cost of central station electricity with a variety of technical, political, and marketing strategies. He replaced reciprocating steam engines with newly developed, more efficient steam turbines, and he ordered them larger to take advantage of economies of scale. He adopted AC to distribute their output to distant substations and users. He sought a large, diverse customer base and adjusted prices to smooth out demand for electricity over the day. This reduced the idle time of generators required for hours of peak demand. By keeping these expensive machines usually running (raising their "load factor"), capital cost per unit of electricity came down, lowering the rate for consumers, increasing their number and the profit for investors.

Insull's Commonwealth Edison Company promoted the use of electricity by vigorous advertising and salesmanship, opening shops to show domestic appliances and motor-driven machinery such as lathes and drills. Insull offered Chicagoans free installation of six outlets to start electrifying their homes. He gave away 10,000 General Electric irons to introduce housewives to this labor-saving device. Commonwealth Edison and Insull's "gospel of consumption" became a model for other urban utilities (Hughes 1983; Platt 1991).

World War I brought coal shortages, unfamiliar in the US with its chronic oversupply of cheap coal. Faced with higher cost and uncertain shipment of coal, companies that had run their own steam plants turned to utility companies for electricity to power manufacturing, food processing, and refrigeration. Central station electricity was more reliable because the large utilities had high priority for coal delivery, and some like Commonwealth Edison formed partnerships with coal companies. Utilities used large turbogenerators that were more efficient than small on-site generators, saving fuel and lowering rates. Central station electricity was now cheaper for a manufacturing plant than its own power production (Platt 1991).

In the decade after World War I, the conversion of American industry to the electric-powered motor created a new economy based on assembly-line methods, standardization, large-scale operations, higher wages, and an outpouring of products for mass consumption. Cars, highways, and power lines facilitated the spread of suburbs. Radio and movies brought mass entertainment. Ever more electricity and oil products fueled the new "machine age." Mechanical servants could remove human drudgery. According to a Commonwealth Edison advertisement in 1925, "The home of the future will lay all of its tiresome, routine burdens on the shoulders of electrical machines, freeing mothers for their real work, which is motherhood" (Platt 1991: 237). War-torn Europe was reconstructing along similar lines.

Regional monopoly utilities, privately owned but regulated by the state, became the norm in the United States though not in all industrial nations, some depending on government-owned electricity providers. Transmission of electricity hundreds of miles over 100 kilovolt lines was by then economical, enabling the integration of coal and hydroelectric generators in the same network, so that one rationally complemented the other – for example, coal carrying the burden during winter and hydro during the spring thaw. Developments in electric supply systems of the 1920s are comparable to developments in railway systems in the second half of the nineteenth century, when major railroads were interconnected, rationalized, and standardized with respect to gauge and equipment. American electrical utilities reached common standards of 120 volts at 60 cycles per second, although there was considerable difficulty keeping connected facilities synchronized at the targeted 60 hertz.

Also like the railroads, there were opportunities for leveraged investors to take control of smaller companies and sell shares in new holding companies. This raised capital for improvements in newly acquired utilities, but it was mostly a money-making scheme. Soon there were "super" holding companies acquiring other holding companies. Samuel Insull was director of one the largest, controlling utilities across the country that together generated one-eighth of the nation's total output of electricity.

For a utility company operating in isolation, provision and operation of peak load generators were expensive. But if peak loads could be smoothed out by pooling operations with nearby utilities having differing load curves, then more demand could be satisfied with the most efficient generators, reducing the need for expensive peaking generators. Also, a pool's combined generating capacity could provide backup when any one utility took a generator offline for maintenance. The first power pool on a large scale, the PJM (originally named PNJ) Interconnection, was formed in 1927 by three IOUs in Pennsylvania and New Jersey, creating what was then the world's largest centrally coordinated grid (Singer 1988). A main PJM dispatcher in Philadelphia requested the dispatchers of each member utility to increase or decrease transfers of power, to optimize overall performance, though it was up to each company to decide how or if it would meet these requests. Managers and operators began to see the interconnection as electrically one company, with a committee of peers from the three IOUs negotiating planning and operations, and forecasting loads. By pooling, the three companies enjoyed economic benefits of a large system but preserved their corporate identities (Hughes 1983). The pool continually expanded to include other IOUs and during the 1930s was a model for pools elsewhere.

There had for decades been large electricity-producing dams at Niagara Falls and in California's Sierra Nevada when the federal government began its major dam building, which slightly preceded the Depression but was considerably reinforced by the need to increase employment through investment in modern infrastructure. The Tennessee Valley Authority, created by Congress in 1933, was a model for giant hydroelectric projects on the Colorado, Missouri, and Columbia rivers. This was the period of engineering "marvels" at Hoover Dam (completed 1936), Bonneville (1937), Fort Peck (1940), and the largest of all, Grand Coulee (1942). Beside electricity and flood control, these created thousands of jobs and irrigation for agriculture. The Federal Power Commission (FPC), originally created in 1920, was strengthened during the Roosevelt administration to license hydroelectric projects on the land or navigable water owned by the federal government. These dams, typically distant from major load centers, required high-voltage transmission lines for their output.

In the 1930s, as the Depression deepened, some of the holding companies collapsed. Insull was indicted in 1932 for securities fraud. Though finally acquitted, his reputation was destroyed, his wealth and health gone. In 1935 Congress broke up holding companies that controlled widely separated utilities but allowed those controlling geographically contiguous companies to continue on the rationale that these could be operationally pooled, smoothing load curves, sharing generating capacity, and benefitting from economies of large-scale operation. The giant American Electric Power Company (AEP) survived the congressional reforms, its monopoly stretching continuously from Ohio to western Virginia.

World War II finally brought America out of the Depression by pushing the economy to maximum capacity. In the early days of the war, as America was furiously increasing production of weapons and materiel, there was a crucial need for aluminum. Arkansas held the largest commercially exploitable bauxite deposit at the time, and the region's power utilities pooled their generation to ensure reliability and dependability during the war and continued their pool afterward. TVA added coal plants to provide electricity for the war effort, supplying the world's largest aluminum plant and later the secret atomic project at Oak Ridge.

Engineering changed during World War II with spectacular developments in radar, communications and atomic energy. After the war, this was reflected in engineering schools, where electric power majors virtually disappeared, with most students going into electronics, computers, nuclear engineering, and aerospace. The innovative brilliance of the electrical industry's first half century was now dimly focused on enlarging generators and extending transmission networks to meet continually rising consumption. More exciting exceptions were the postwar promise of cheap electricity from the atom, and the development of computerized systems for controlling power networks. The introduction of computerized control in the late 1950s improved analysis of power flow and transient stability, and about the same time techniques were introduced to predict probabilities of generation adequacy.

The Atomic Energy Commission (AEC) was established in 1946 to transfer control of nuclear energy from military to civilian hands. The manufacturing approach was essentially to turn the reactor that functioned well in nuclear submarines into the heat source for steam-turbine generated electricity, housing it in a steel-reinforced concrete containment in case of radiation leakage. Westinghouse and General Electric offered "turnkey" projects at cut rates to encourage their adoption by utility companies. By the 1960s utilities turned enthusiastically to nuclear power plants.

Since an IOU's profit was, by state regulation, a fixed percentage of its capital investment, an increase in investment (often in nuclear plants) increased profit. Investment was justified to the regulatory agency by increasing public demand for electricity, which the company (and associated industries) promoted. These circular incentives encouraged continued increase in the use of electricity. Kilowatt-hours became cheaper, and their consumption grew in both absolute terms and as a proportion of total energy consumed. During the 1960s annual growth rate of electricity usage reached seven to nine percent. As neighboring utilities interconnected, they enjoyed the benefits of sharing generation reserves and averaging load curves. This opened new opportunities for each utility to sell its excess electricity to another. By the 1950s, this was not unusual (Beck 1988). The utility business was growing smoothly, and profits seemed assured. IOUs were bastions of modernity and economic stability, safe places for "widows and orphans" to invest money.

Things fall apart: 1965+

On November 9, 1965, a massive power failure cascaded across the northeastern United States and Ontario, affecting 30 million people. Though it would be surpassed by the catastrophic power failure of 2003, affecting about 55 million people in eight states and two Canadian provinces, the 1965 blackout was a bigger shock to the national psyche and a complacent electric power industry. "The initial reaction . . . was one of general disbelief that such an incident could happen" (FPC 1967, Vol. 1: 1).

The proximate cause of the 1965 blackout was incorrect relay settings on five 230-kilovolt transmission lines between Niagara Falls and Toronto; this caused overloading of other lines and a reversal of the power flow, which surged into the United States, tripping out circuit breakers across the Northeast. Power failures were not unusual, but this was far bigger than ever before. After the most extensive investigation of the nation's power system in its 85-year history, the FPC's recommendation was more and stronger interconnection.

At the time of the blackout, reliability criteria were the business of individual systems. Another FPC recommendation was the development of more uniform (though still noncompulsory) reliability standards. By 1968, nine regional reliability councils had formed, covering the lower 48 states and Canada, and joined together in voluntary association as the North American Electric Reliability Corporation (NERC) to improve the dependability of the nation's electricity supply. It became obvious with the great blackout of 2003 that this was an insufficient arrangement. (In 2007 the US Federal Energy Regulatory Commission, successor to the FPC, granted NERC the legal authority to enforce reliability standards on all users, owners, and operators of the bulk power system in the United States.)

In retrospect, we see more signals during the 1960s that the "maturity" of the electric utility industry was a misnomer. Utilities were focused on intense marketing, placing orders for huge new generators to meet need. Manufacturers responded by making bigger versions of the same designs that in prior decades had improved economies of scale with greater thermal efficiencies. But by the 1960s thermal efficiencies had leveled off, and some scaled-up units had defects in manufacture and problems in operation. Consolidated Edison's "Big Allis," named for its manufacturer, Allis Chalmers, was put on line in 1965, the first steam generator of 1,000 megawatts. When it short-circuited in 1970, New York City lost 14 percent of its power supply. Engineers worked for six months to bring Allis back to service. Then, after 87 minutes of generation, Allis conked out again and took another four months to return on line. Thirty-nine days later she tripped once more, this time because of faulty bearings (*Time* 1971). Machines so large and complex are difficult to repair, and when one is off line it leaves a big hole in the electricity supply. "Economies of scale" had reached their limit.

Nuclear power plants were scaled up too, from 600 megawatts in 1965 to 1,152 megawatts in 1973. Designed for individual sites without standardization, they suffered construction and regulatory problems. By 1974 the lead time for fossil fuel units was eight years, for nuclear units ten years, and costs were escalating (Hirsh 1989). Believing that the likelihood of a serious accident was nil, industry attitudes about nuclear safety were optimistic. Consolidated Edison proposed in 1963 to site its Ravenswood nuclear plant in the heart of New York City, a plan later abandoned. Estimates of the cost of constructing nuclear plants, of their immunity from accidents, and the ease of disposing of spent fuel would all prove to be overly optimistic (www.nrc.gov/about-nrc/short-history.html). By the late 1960s, electricity rates were going up and state regulators, once compliant to utility requests, became resistant to rate increases. Before 1969, no state had a law requiring site approval for a new plant, but 21 states had such laws by 1976 (Hirsh 1989).

Each year brought new problems. The Clean Air Act of 1970 (strengthened in 1990) mandated reductions in sulfur dioxide, requiring that many fossil fuel power plants either install scrubbers to clean SO_2 from stack gases or switch to low sulfur fuel – both expensive. By 1971, fuel shortages and construction delays caused brownouts. The Yom Kippur War of 1973 triggered OPEC's oil embargo, escalating the price of oil (and other fossil fuels) and initiating the "energy crisis." In 1974, New York's Consolidated Edison missed a quarterly dividend, the first time in decades. In 1977 President Jimmy Carter named Defense Secretary James Schlesinger as his "Energy Czar" and asked Congress to create the Department of Energy. The Iranian revolution of 1979 pushed petroleum prices far higher, essentially ending oil as a fuel for electricity generation.

Despite oil shortages, the 1970s were a period of strong public opposition to nuclear power, climaxed by the 1979 accident at Three Mile Island. The Nuclear Regulatory Commission (NRC), which inherited the regulatory function of the old AEC, required retrofitting of nuclear plants to conform to new safety standards, increasing delays and costs of construction. The combination of public complaint, rising expense, and the decline of load growth as the price of electricity went up, ended contracts for new nuclear power facilities in the US. The Shoreham plant on Long Island, New York, newly completed at a cost of $6 billion, was closed by protests in 1989 and decommissioned without generating any commercial electrical power; another nuclear plant in California, already operating, was closed by public vote the same year. Coal, which had been suffering long decline as a dirty, polluting source of energy, was resurrected as the cheapest and most plentiful fuel for electric generators. (Global warming, with its implication for coal burning, did not enter the public arena until 1988.)

Attempting to supplement the nation's electricity supply, Congress in 1978 passed the Public Utilities Regulatory Policies Act (PURPA), which encouraged nonutility companies to use "alternate" means (cogeneration and renewables) to generate electricity. PURPA required IOUs to buy this electricity at guaranteed favorable prices and to carry it on their transmission lines. The Act exempted qualified cogeneration and small power producers from most state and federal regulations, forming a class of power companies that existed outside the realm of normal utility operations. This was the first serious breach of Insull's model of the electric utility as a natural monopoly in its area of operation, the first step toward deregulating the industry.

The Energy Policy Act of 1992 created another new class of power producers, called "exempt wholesale generators," which could be owned by a utility, a utility holding company, or a nonutility developer. The 1992 law requires IOUs to open their transmission lines to any exempt wholesale generator. For example, a nonutility developer could build a gas turbine generator, sell its output to a wholesale customer at a price lower than asked by an IOU, and then (for a fee) wheel the electricity over the IOU's transmission lines to the customer. These new rules allowed energy traders such as the Enron Corporation to make (and manipulate) electricity markets, rather like stock or commodity exchanges. The amount of electricity sold wholesale by nonutility companies grew from 40 to 222 million megawatt-hours between 1986 and 1995. Wheeling increased at an average annual rate of seven percent over this period (Warkentin 1998).

States differed in their enthusiasm for deregulation. California was especially generous to nonutility generators and electricity traders, but created an inconsistent system in which retail sales of electricity remained regulated while wholesale prices were set by the market. Enron and other smart traders gamed the rules, driving prices higher (McLean & Elkind 2003). In 2000 and 2001, California's IOUs had to buy electricity at far higher prices than they could sell it for. The state suffered rolling blackouts and brownouts, and its electricity market was in chaos. The Pacific Gas and Electric Company, one of California's major IOUs, declared bankruptcy, and another, Southern California Edison, barely avoided it. The California debacle led other states to put their plans for deregulation on hold, a caution later reinforced by the Great Recession of 2008.

Electrification in Japan

The Chicago Fire of 1871 and the San Francisco earthquake of 1906 obliterated major portions of these cities, giving planners a clean slate upon which to rebuild more rationally than before, reconfiguring streets in Cartesian grids and adopting building codes that limit vulnerability to fires and earthquakes. The industrial infrastructures of Germany and Japan were destroyed in World War II, offering similar opportunities. However cultural traditions persist, so there is never a really clean slate, wholly divorced from the period before destruction. Surviving images of how things *were* often guide how things *should be* in the future, one reason that reconstruction can occur very quickly. Postwar Germany, its architectural landmarks destroyed, set builders to working from old blueprints and pictures to meticulously build again the "historical" buildings now gone. A visitor can scarcely tell whether an "old" church or *Rathaus* is original or a reproduction.

As it had during the Meiji Restoration, Japan in the late nineteenth century looked to Europe and America for its modernizations. Tokyo Electric Light Company was founded in 1883, the year after Thomas Edison opened his Pearl Street generating station in Manhattan. In 1895 Tokyo's entrepreneurs installed 50-hertz AC generators manufactured by the German corporation AEG. The next year another new company in Osaka purchased 60-hertz AC generators from General Electric in the United States. There was no thought at the time that the two cities, 500 kilometers apart, might pool operations.

By 1945 there had been considerable consolidation of Japan's local grids, though as elsewhere, they were small and poorly connected by today's standards. That infrastructure was demolished by American bombers, giving the planners of Japan's remarkably successful reconstruction the opportunity to start their electricity system from scratch. What they constructed was modern by the standards of postwar America, using nuclear and fossil fuel power plants, and eventually tying them together with high-voltage transmission lines. By 2011, Japan ranked third in the world in total electricity production (after the US and China) and 18th in per capita electricity consumption. Nuclear, coal, and natural gas each produced about a quarter of Japan's electricity, with oil and hydroelectricity each supplying about eight percent.

By that time Japan had an extensive grid system of very high-voltage transmission lines (Figure 8.1). But oddly, in rebuilding the system, planners retained the old AC frequencies: 50 hertz in Tokyo, 60 hertz in Osaka. It is unclear why they did not settle on a single frequency. Perhaps the bombardment left undamaged generators that could be restored to service, or perhaps it was a remnant of the traditional past. In any case, as the grid – actually two grids – expanded, Tokyo and the rest of eastern Japan ran on 50-hertz electricity, while the big cities and countryside southwest of Tokyo ran on 60 hertz. Because the two grids are not in synchrony, AC electricity cannot directly flow between the eastern grid and the western grid.

Tokyo and the Fukushima Daiichi nuclear complex are on the eastern grid. With Fukushima's reactors disabled by the 2011 earthquake and tsunami, and other reactors shut down as a precaution, the eastern grid was starved of power. The nation called for and obtained voluntary conservation. Still, Tokyo Electric Power Company (TEPCO), operator of the Fukushima complex and power supplier to greater Tokyo, had to ration electricity by instituting rolling blackouts; power would be on for only portions of the day.

The western grid, largely unscathed by the earthquake and tsunami, still had sufficient generation capacity to cover TEPCO's peak demand if the electricity could be transferred between grids. There are three small facilities that do connect the grids by first converting AC from the sending grid to high-voltage DC, then converting that back to AC in synchrony with the receiving grid. Together the three connecters can transfer about one gigawatt of capacity, barely touching the 9.7 gigawatts lost with 11 nuclear reactors offline.

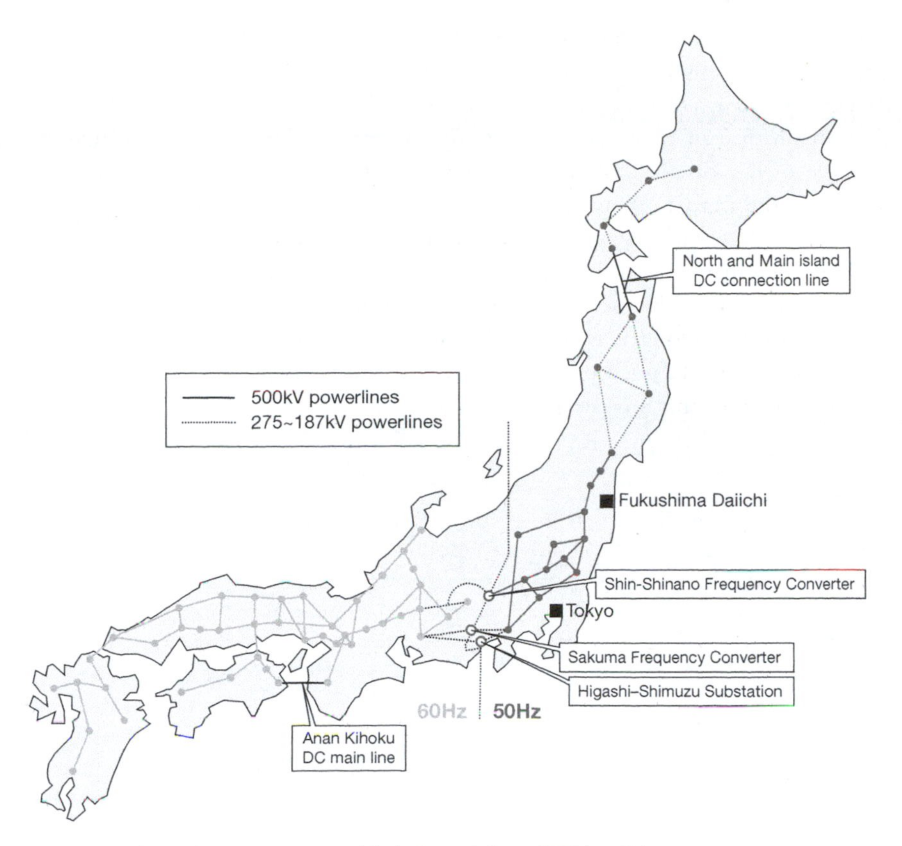

FIGURE 8.1 Japan's two power grids (adapted from Wikipedia)

Japan's nuclear and electricity crises are ongoing. Almost certainly it will return to considerable nuclear generation because there is no realistic option in the short term. The nation recently announced a commitment to renewable energy. If the Japanese devote their famous ingenuity to creating a truly modern electrical system, they may become a model for the rest of the world.

Conclusion

The industrial world depends on electric grids that – for all their advantages – frequently fail in minor ways and occasionally in disaster. Still, there seems no realistic option to do without them. Proposals for fully dispersed electricity generation, with each building or community an island providing its own power, ignore the advantage if not necessity of interconnection with distant generators to make up for local deficits or outages. If we hope to shift our mix of energy sources away from fossil fuels and toward renewables, grids must become even more expansive and interconnected because the best regions for solar, wind, hydroelectricity, and geothermal generation are usually distant from the major load centers and must be connected to them with new high-voltage transmission lines.

Japan's case is especially interesting because the destruction of World War II gave that nation an opportunity to construct a "modern" grid, leaving behind the baggage of the first half century of blind growth, yet still the planners did not anticipate the need to transfer electricity between the eastern and western grids. No one would wish for so painful an opportunity again. Improvements must instead be built incrementally on the massive systems that are already in place. Electrical generation is one of the most capital-intensive industries in any developed nation, so substantially altering it is no small feat. To understand how grids might be improved, we need a good picture of how they got to their present state. For this we will return to the American electrical system.

9

AMERICA'S THREE GRIDS

In America and other industrial nations that escaped the destruction of World War II, electrification evolved after Edison's time without traumatic interruption. While new elements were always added to existing infrastructure, and traditional practices were modified, the system never could – and probably never will – be wholly rejuvenated from scratch.

Today the United States (excluding Alaska and Hawaii) and Canada receive most of their electricity from four regional grids: the Eastern Interconnection, the Western Interconnection, the Texas Interconnection, and the Quebec Interconnection (Figure 9.1). My focus is on the three grids that cover the US: East, West, and Texas (often called "ERCOT," the acronym for Electric Reliability Council of Texas). The Eastern Interconnection essentially corresponds to the bright and more densely populated eastern half of the United States as seen at night in the cover illustration.

When (and why) did it become clear that the nation would today be covered by three essentially separate grids? To what extent was this tripartite outcome due to coordinated planning since World War II, to what extent happenstance?

The three grids differ considerably in size (Table 9.1). The elements of each regional grid are synchronized to their own 60-hertz alternating current, but are not synchronized with the other regional grids.

How did the three grids grow after World War II?

America's three-grid structure did not exist nor was it foreseen in the decades immediately after World War II, a period when electricity consumption was growing at seven to nine percent annually. But regional grids were growing nearly everywhere. The most active period for constructing the highest-voltage transmission lines (i.e., 345 kilovolt and above) occurred between 1960 and 1990

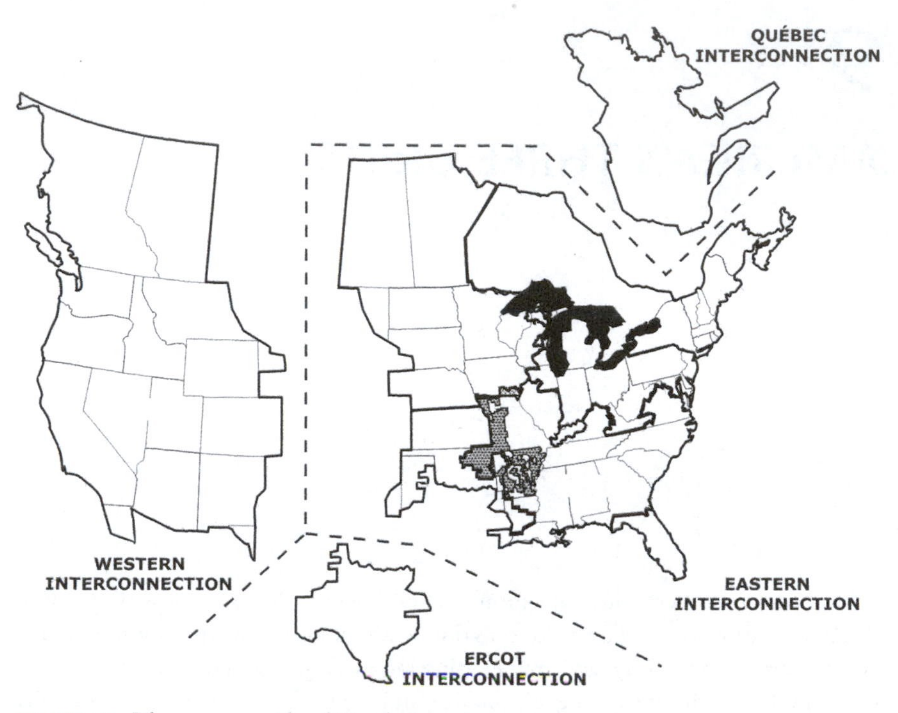

FIGURE 9.1 The power grids of North America

TABLE 9.1 Sizes of the three grids in 2008, excluding Canada

	Eastern Interconnect	*Western Interconnect*	*Texas (ERCOT)*
Number of customers	101 million	66 million	22 million
Net electricity generated (gigawatt-hours/year)	2,941,000	755,000	405,000
Installed generation (megawatts)	795,800	190,000	84,000
230+ lines (km)	150,260	93,300	14,270

Sources: State-level data reported by the U.S. EIA and aggregate data supplied by grid associations

(Figure 9.2). There is some irony in this timing because it coincides with the transformation of the US electric system from a period of untroubled growth – some called it "maturity" – to one of uncertainty, even chaos. Some of the planning of these lines occurred in the halcyon days before 1965; some was in response to the Northeast blackout of that year. By the 1990s, high costs and other problems afflicted the industry, and restructuring had shifted major profit-making opportunities to generation and marketing, not transmission. Line construction tapered off, even as demand for electricity continued to rise.

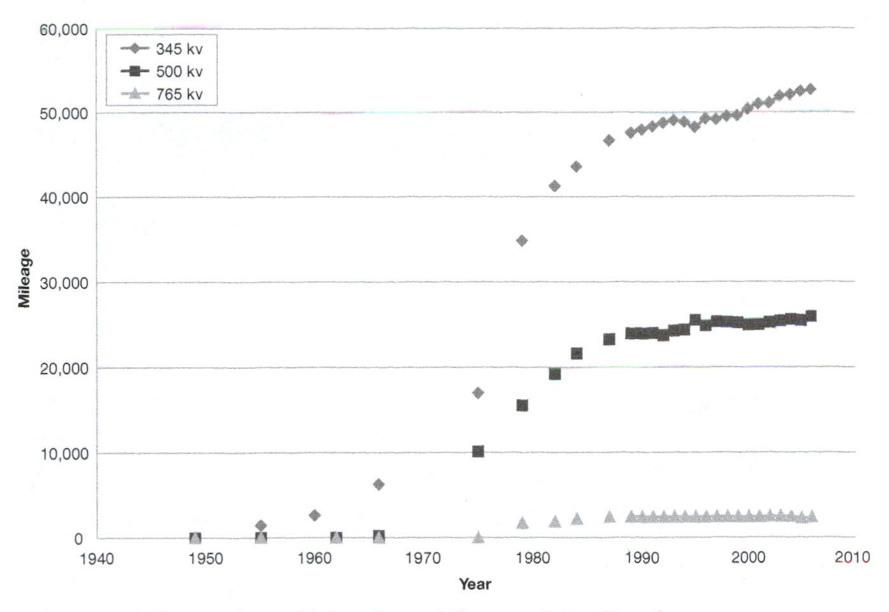

FIGURE 9.2 Mileage of very high-voltage AC transmission lines by year

Mapping the grid

To trace the geographical spread of the grid since World War II, I located nine maps of high-voltage transmission lines, published between 1949 and 2010. The Department of Energy (or the earlier Federal Power Commission) produced maps from 1949 through 1978. PennWell Publishing Company sold a map for 1994, and Platts (McGraw-Hill) sold one for 2010. Figure 9.3 shows maps for 1949, 1962, 1978, 1994, and 2010, converted to comparable formats. Not all published maps show interconnections to Canada, so these are ignored here, an important omission.

The map for 1949 shows operating lines of about 230 kilovolts, the highest voltage then in service. These were in five areas of the nation, four of them based on large hydroelectric projects – in California, Pennsylvania, Washington, and New Hampshire.

Southern California Edison Company placed the first of these lines in service in 1923, bringing hydroelectricity run by snowmelt in the Sierra Nevada Mountains to Los Angeles, much of it to electrify trolleys. The mountains north of California's Central Valley were another source of snowmelt, so in 1923 the Pacific Gas and Electric Company ran a 220 kilovolt line from Mount Shasta to a substation near Sacramento (EEI 1962). In the early 1930s, the US Bureau of Reclamation ran a 287 kilovolt line from Hoover Dam (in south Nevada, on the Colorado River) to Los Angeles. In 1945 the Bureau of Reclamation finished Shasta Dam, the second largest dam in the US, its electricity routed southward toward San Francisco. By 1949, California had more interconnected very high-voltage transmission lines than any other region of the nation.

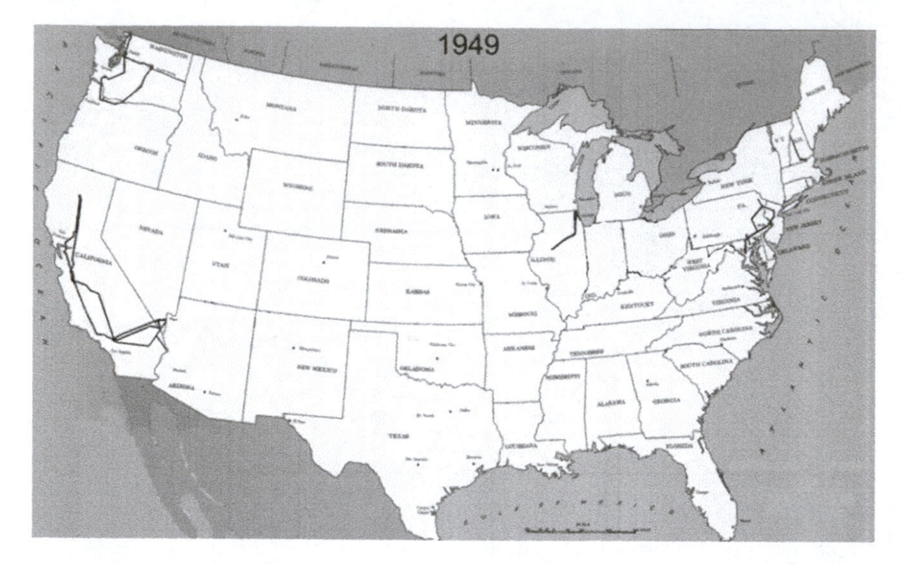

FIGURE 9.3 A–E Growth of the US power grids, 1949–2010

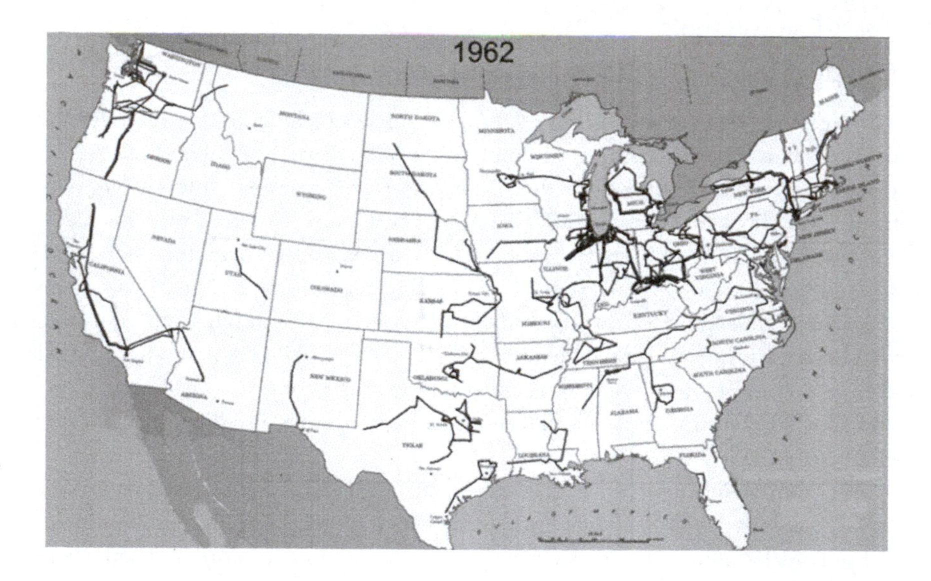

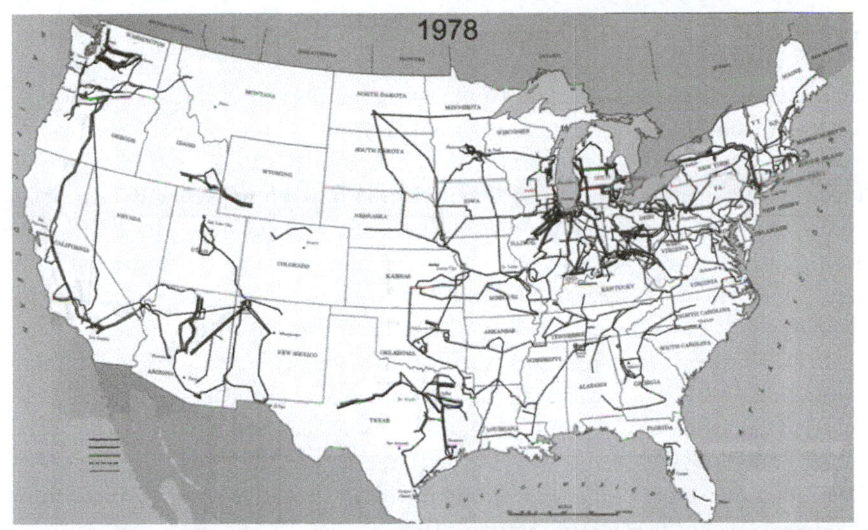
1978

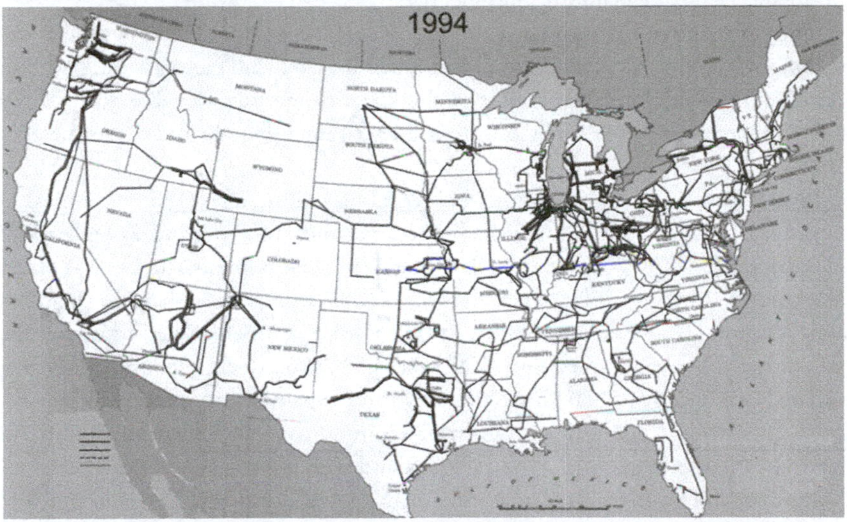
1994

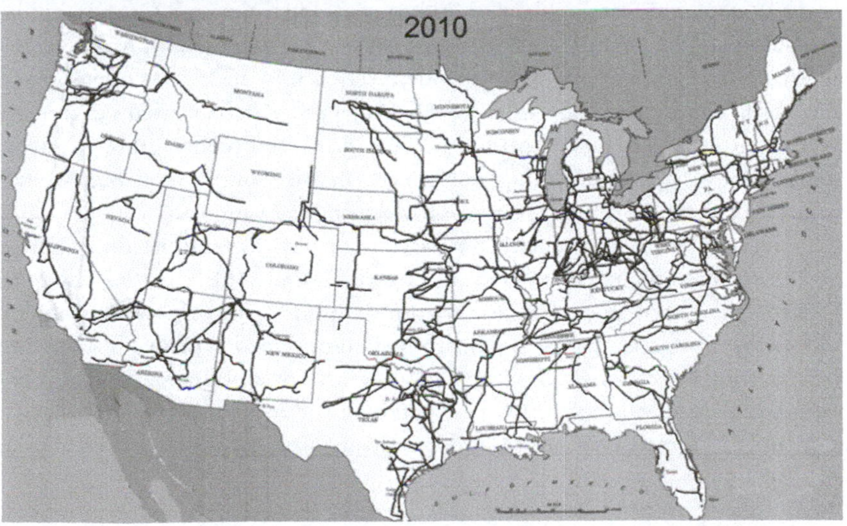
2010

In 1928, Philadelphia Electric Company, a member of the PJM pool, opened a 220 kilovolt line to bring in power from the lower Susquehanna River, at that time the largest hydro development in the world. By 1932, 220 kilovolt lines made a ring configuration over Pennsylvania and New Jersey (Chase 1928; http://americanhistory.si.edu/powering/transmit/gallry26.htm). In 1930, New Hampshire's first 220 kilovolt line began carrying electricity from generators on the Connecticut River, at the time New England's largest hydroelectric development, 240 kilometers (150 miles) south to Boston (EEI 1962).

In Washington State, from 1939 to 1941, the Bonneville Power Administration created a loop of 230 kilovolt wires linking its two great new dams on the Columbia River, Bonneville and Grand Coulee (the nation's largest), to loads in Portland, Puget Sound, and Spokane. World War II accelerated the need for electricity at Washington's shipbuilding and aluminum plants, and the secret plutonium-producing facility at Hanford (Springer 1976).

The only one of these early nuclei that was not based on hydroelectricity was in Illinois. Commonwealth Edison, while still under Insull's command, had built a generating complex in the center of the state, near easily minable coal, its first unit completed in 1928, sending output 230 kilometers (145 miles) northeast to Chicago (EEI 1962).

In the early 1960s the nation had 265,000 kilometers (165,000 miles) of transmission lines of 110 kilovolt or higher, but only four percent as high as 345 kilovolts (FPC 1967, Volume 2: 3). The first 345 kilovolt line in the US was placed in service by American Electric Power in 1953, linking power plants in Ohio and West Virginia. (A 345 kilovolt line cost three times as much as a 115 kilovolt line but carried about nine time as much power [EEI 1962; Beck 1988: 86].) My map for 1962, still mostly of 230 kilovolt lines, shows expansion of the five nuclei since 1949. The densest connections ran from Commonwealth Edison in Illinois and other Great Lakes states – the industrial heartland with good access to coal – eastward to the coast and southward to the Tennessee Valley Authority (TVA). On the West Coast, Washington State's enlarging grid moved down through Oregon, aiming for California. In California itself, the network of operating 230 kilovolt lines had not changed much since 1949. Independent Texas now had its own big lines. A new north-south connection from North Dakota to Kansas carried power from the Bureau of Reclamation's Missouri River dams, the highest (though not an especially large generator) being Fort Peck. In the Southeast, a high-voltage corridor – mostly TVA lines – ran from Mississippi to Virginia.

Despite this expansion, today's three-grid structure was not yet discernible in 1962. A report published by the Edison Electric Institute (EEI) in 1962 has a map of "principle interconnected electric transmission lines" (mostly of relatively low voltage) for 1960, their degree of synchronicity unspecified. The map shows many lines covering the eastern half of the nation, a sparseness of lines across the Great Plains and Rocky Mountains, and concentrations of lines in the Columbia River Basin and California. Presence of lines essentially corresponds to the distribution of population.

Historical overviews during the 1960s describe the nation's power system as divided into six major operating groups (EEI 1962; Moorehouse 1965; Friedlander 1966). Construction of a number of ties between the PJM pool and neighboring networks in about 1962 was effectively the beginning of today's Eastern Interconnection.

In 1967 the Federal Power Commission, responding to the Northeast blackout of 1965, performed an extensive National Power Survey of the US electric system and gave a perhaps exaggerated overview of its integration:

> The framework of interconnections [among utilities] has changed materially since 1939 when the electric power industry of the United States was divided into numerous relatively small groups of interconnected systems. The progression of interconnections from 1939 to today's [1967's] pattern [produced] nearly complete interconnection throughout the country . . . Although *the great majority of the nation's power systems are now interconnected in a single network*, the network is not strong enough at many points to assure adequate support in the event of emergency.
>
> (1967, Volume 1: 34, italics added)

Contrary to its text, the FPC maps showed Texas as insular in 1967 as it was in 1939 (and nearly is in 2012). An FPC map for 1967 shows the nation almost completely interconnected, surely an overstatement. The accompanying text lists several lacuna; for example, the Northwest was not connected to the Southwest except indirectly through the mountain states via lines of "very limited strength," or Michigan was not connected to the rest of the US except through Ontario. Overall, the interconnection that did exist was of low capacity and not generally synchronous.

It is important to distinguish the physical networks of wires from the social/organizational networks of IOUs and public agencies that cooperated in pooling their generation and transmission facilities, and in wheeling electricity. The FPC's 1967 report contains, besides its mapping of the physical networks, another map showing 18 major power pools, each a formal organization comprising neighboring utilities (usually IOUs) that more or less shared power. These pools were modeled to varying degrees on the PNJ interconnection, where IOUs had cooperated successfully since 1927. Pooling was now growing, the pools continually reconfiguring as their span of cooperation increased.

In a tidy world, the physical and organizational networks would coincide. This was true in the West. The pool called the "Northwest Coordination Group" encompassed the physical nucleus of transmission lines based on Columbia River hydro; the California Power Pool matched California's network of big lines. But in the East there was no one-to-one relationship between an organizational pool and a physical network of wires. It is worth noting, however, that the 16 power pools (plus TVA) that covered the eastern half of the nation in 1967 correspond to what would become the Eastern Interconnection.

In the 1970s and 1980s there was rapid increase in mileage of 500 and 765 kilovolt lines (Figure 9.2). My maps for 1978 and later do not show lines smaller than 345 kilovolts. By excluding these now "small" lines, the later maps understate the degree of interconnection across the nation, but in my judgment they more accurately reflect the changing geographical pattern of fully coordinated connectivity.

By 1978 the present three-grid structure is nearly realized in the network of very high-voltage lines (Figure 9.3c). There is a new connection between previously separate physical nuclei in Washington State and California (Dominy 1969). Los Angeles's link to Hoover Dam now continues eastward through Arizona to New Mexico and then northward into Utah. The major node in this extension is the huge Four Corners Power Plant in the northwest corner of New Mexico, fueled by subbituminous coal in large deposits nearby, its first unit online in 1963. Up north, further damming in the Columbia River Basin brought big transmission lines from southern Idaho to Wyoming.

By now the nation was perceived as being covered by three major, essentially separate grids. The earliest description I know in terms of the Eastern, Western, and Texas Interconnections is an Associated Press article by John Cunniff appearing July 15, 1977 and sourcing an FPC official.

Muscular Texas had a serious network of large lines by 1978 that easily could have been tied to the eastern grid but instead the state maintained its separation. Coordination of utility operations in Texas began during World War II when excess generation was sent to the Gulf Coast to support the war effort. After the war, most of these utilities, covering about three-fourths of the state's land area, realized the advantages of remaining interconnected and in 1981 transferred all operating functions to the Electric Reliability Council of Texas (ERCOT).

The Rocky Mountains may be imagined as a separator of east and west, but they were not particularly daunting as an altitude barrier. More important is the sparseness of population (i.e., few customers) in the Great Plains and High Plains, and few large generators or loads across that expanse. To keep distant parts of an AC network in synchrony, large nodes must be fairly closely spaced. For this reason, if no other, it was impractical to synchronously integrate the eastern and western grids.

Construction of large transmission lines nearly leveled off in the 1990s. My map for 1994 (Figure 9.3d) shows a vivid separation between the West and the eastern half of the nation. On either side of this divide there is increased interconnection. Among the western states, for example, the Idaho lines now extend westward and southward. The eastern grid now runs all the way to Florida. Not shown are important connections to Canada. My 2010 map (Figure 9.3e) shows still more density of lines but the geographical distribution is not fundamentally different.

Connecting the three grids

There is a slight fiction in portraying the three US grids as separate entities. More accurately, each grid is an internally synchronized 60-hertz network, out of synchronization with the other grids. It is possible to send some electricity from one grid to another. This is accomplished at junction points where, as in Japan, AC from the sending grid is converted to DC, which is then converted back to AC in phase with the receiving grid. One connecting line, established in 1988 through Nebraska's Virginia Smith junction, was large enough to be visible on my 1994 and 2010 maps.

At quick glance, Texas looks like part of the Eastern Interconnection, and it easily could be, but the Texas legislature and utilities have continually protected their independence from federal interstate regulation. Two DC ties to Oklahoma, established in the 1980s, link ERCOT to the Eastern Interconnection, together able to transfer about 785 megawatts, equivalent to a large power plant. Texas has managed through legislation and court decisions to hold these connections as jurisdictionally inconsequential because they do not connect the AC grids synchronously, so ERCOT maintains its status as an independent grid (FERC 2000; Fleisher 2008). Similar standards have been applied to the physical junctions between the other grids.

Unsynchronized grids can also be connected with variable frequency transformers (VFTs), which allow a controlled flow of electricity while isolating the independent AC frequencies on each side of the junction. Today the Eastern Interconnection is tied to the Western Interconnection with six DC ties in the US and another through Canada, to the Texas Interconnection with two DC ties, and to the Québec Interconnection with four DC ties and a VFT. DC junctions help insulate an area from propagation of cascading outages but do not wholly prevent them. There is no direct tie between the Western Interconnection and ERCOT. Usually electricity transfers between grids are modest but not trivial. Since the 1990s, the flow of hydroelectricity from the huge James Bay Project in Québec to the Eastern Interconnection has become considerable, through junctions as large as 2,000–3,000 megawatts. However all of these grid-to-grid connections, whether international or through US junctions, are small compared to the indigenous generation capacity of each grid. Even taken together, they do not significantly distort the essential characterization of the US as being serviced by three distinct grids.

Present outlook

Beginning in the 1990s, several states deregulated electric utilities, requiring owners of transmission lines to allow independent private generators to send their electrical output over the grid and sometimes requiring utilities to divest themselves of their own generators. Increasingly, the business of generation was split from the business of owning and maintaining the network of wires. The generation or brokering of electricity offered high potential for profit-making and attracted the most investment, with relatively little money devoted to maintaining or upgrading physical lines. As a result, American transmission networks are generally regarded as congested, needing repair, replacement or augmentation, and requiring new technology to improve future operation (Casazza 2007).

Some have suggested that a major expansion and upgrading of the US electrical system is needed, possibly integrating the regional grids into one national network and introducing "smart grid" technology, which accepts feedback from consumers to improve pricing and usage efficiencies (Fox-Penner 2010). The prospect of replacing fossil fuel-generated electricity with generation from wind (especially in the belt of states from North Dakota to Texas, or off the ocean coasts) and solar (especially from the desert Southwest) requires new transmission lines to major load centers. Very long transmission routes would be required to supply East Coast users with electricity from the major renewable sites in the western half of the nation. Expansion of the grid would also facilitate the wheeling (sale) of electricity between distant points.

Solar electricity from the Southwest is intermittent in a mostly predicable way, rising and falling with the sun, so it can be smoothly integrated into the grid at low levels but can provide challenges when cloud cover changes quickly. Wind is chaotic, the power generated by wind turbines capable of unpredictable change within minutes (Charles 2009). Since air currents vary from one locale to another, wind input can be smoothed out by aggregating many windmills that are spread over a wide geographic area, but this requires considerable extension of transmission lines as well as flexible backup generation.

Few would question the desirability of improving the grid, but it is not certain that this is best accomplished by making it bigger and more interconnected. The most obvious objection is that a very large integrated grid is vulnerable to widespread system failure reminiscent of the great 2003 blackout in the Eastern Interconnection, when about 55 million Canadians and Americans lost power (Park 2010). Larger grids open the possibility of even bigger outages because no complex, highly interconnected system is failure-proof (Perrow 1999). A report of the Electric Power Research Institute emphasizes that a sudden and unplanned change in conditions at one location in the grid can have immediate impacts over a wide area, magnifying as it propagates through the network. "Large-scale cascade failures can occur – almost instantaneously – with consequences in remote regions or seemingly unrelated businesses" (Amin 2003). In August 2012, on two successive days, India suffered the largest blackout ever, leaving 600 million people (ten percent of the world's population) without a working grid, despite having a reportedly sophisticated system of circuit breakers. The only consolation was that so many Indians remain unconnected to the grid, depending on either local generators or no electricity at all, that the pattern of life was not greatly disturbed.

Widespread system failure, whether from accident or attack, is potentially catastrophic for an industrial nation. Therefore consideration should be given to a strategy of at least partially disaggregating the US-Canadian electrical system into relatively isolated modules, i.e., smaller and more independent grids, having interconnections that could be used in emergencies. With modular grids the impact of a blackout is intrinsically limited, yet it would still be possible to import electricity should an emergency arise. One proposal is to divide the large grids into ten or 12 smaller areas, tied together with direct current connections, which would partially control cascading failures (Käuferle 1972; Loehr 2007).

Wheeling is another factor that plays into the desire for an extensive versus modular grid system. For profit-making purposes, electricity traders want the ability to buy cheap electricity from one area and sell it at higher price in a distant area; this requires long-distance transmission lines. On the other hand, the farther electricity is sent, the more energy is wasted as "line losses." Also, wheeling electricity from area A, through area B, to area C, increases power line congestion in area B, though B is neither the generator nor consumer of the electricity.

Public and environmental opposition to high-voltage transmission lines is another problem for an extensive grid, certainly for any plan to send wind and solar electricity long distances. Citizen's groups in Minnesota and Upstate New York vociferously opposed the construction of new, very high-voltage power lines in the 1970s and such protests recur to this day (Mazur 1981; Wellstone & Casper 2003). Furthermore, there is some local governmental opposition to an extensive grid, most obviously the desire of Texas to keep its grid a separate entity, thus avoiding federal regulation of interstate transmission.

Efficiency and reliability of the three grids

Despite being functionally isolated, the three US grids are never analyzed as separate, integrated systems, a comment on the disaggregated oversight of continental electricity, even within the boundaries of a single nation. NERC divides the Eastern Interconnection into six reliability regions based on historic associations. Utility companies and transmission organizations focus on their own local operations. Such narrow scope gives short shrift to the pieces as components of larger integrated systems.

Before undertaking a wholesale consolidation of the separate grids, it is worth inquiring if larger grids provide better efficiency and reliability than small grids. The three existing grids allow us to ask if the largest (the Eastern Interconnection) works best, and the smallest (Texas-ERCOT) worst. Based on their experiences, does operation (at least efficiency and reliability) improve with size?

My colleague Todd Metcalfe and I define the *efficiency* of each grid as electricity consumption by its end users divided by its net generation. This definition ignores efficiencies of power plants themselves because net generation is the amount of electricity that is output from a plant, without regard to the plant's efficiency in converting fuel to electricity. Efficiency here represents the portion of electricity entering the grid that is finally sold to end users, the difference being substantial losses due to transmission and distribution, and also smaller losses due to electricity theft and meter inaccuracies. (We adjust for small transfers of electricity between grids.)

We measure a grid's *reliability* by its blackouts or other power interruptions that affect at least 50,000 customers. (Most outages affect fewer customers.) For each large disturbance, we multiply the number of customers who lost power by the reported duration of the outage in hours, the product being customer-hours of outage. To fairly compare the three grids, customer-hours of outage for each grid is divided by that grid's total number of customers. There are trustworthy outage data for all three grids only since 2007, while good data for efficiencies go back to 1990.

Our comparisons show that grid operation, in terms of efficiency and reliability, is not essentially a function of grid size. The huge Eastern Interconnection is no more efficient and is clearly less reliable than the Western Interconnection. On the other hand, the smallest grid, ERCOT, is less efficient than the other two, no more reliable than the East and less reliable than the West. As in the story of Goldilocks, the middle-size grid, the Western Interconnection, is best overall.

There are caveats. Anyone remembering the California electricity crisis of 2000 and 2001, generally blamed on price manipulation and partial deregulation of the electricity market, would be rightfully loath to regard the Western Interconnection as a paragon. Perhaps 2007–10 was an uncharacteristically good stretch for the West. Furthermore, basing our high evaluation on two indicators of performance is surely too limited. It is worth emphasizing again that our measure of efficiency begins with the electricity that leaves the generators; it does not include the efficiency with which the generators convert primary fuels into electricity. Furthermore we cannot distinguish between efficiencies of high-voltage transmission lines and efficiencies of the lower-voltage distribution lines, which under restructuring are often the responsibility of different organizations. We must also bear in mind that systems can often be improved to some extent by spending more on them, so each grid's operation depends in practice on a balance between performance and cost.

Why does ERCOT lag behind the other grids in efficiency? We can eliminate as answers any recent developments such as the inordinately large integration of wind power in Texas, because the difference in efficiencies extends back at least to 1990. It is unlikely that Texas lacks price incentives for efficiency because its retail price for electricity is above the average for all states. Possibly if ERCOT were connected synchronously to the Eastern Interconnect it would be more efficient. Possibly the Texas power industry's aversion to regulation fosters inefficiencies in transmission, though one could argue contrarily that by being unfettered, the Texas grid is more efficient than it would otherwise be.

Why was the Western Interconnection more reliable than the other two grids in 2007–10? One might think that the West's freedom from hurricanes, or the low susceptibility of its population centers to ice storms, could account for its advantage. But in fact there is no seasonal variation in reliability among the grids, rendering explanations based on seasonal weather unlikely. The superior reliability of the Western grid can be partly attributed to faster restoration of power, but this begs the question: Why is restoration of power faster in the West? One contributor is the relatively high incidence of load shedding in the West, which is quickly restored. However, even after excluding incidents of load shedding, outages in the West remain shorter than in the East or Texas. More substantive factors remain to be explored, for example, the possibility of enhanced reliability of transmission and distribution in the West due to undergrounding or other protective measures. In any case, present experience gives no assurance that increasing the size and connectivity of a grid will in itself improve operation (Mazur & Metcalfe 2012).

Conclusion

The US National Academy of Engineering calls electrification the greatest engineering achievement of the twentieth century (http://www.greatachievements.org/), this despite widely perceived inadequacies in the North American power system as it begins its second century of operation. The system must be continually improved to meet ever-increasing demands for electricity, to improve reliability, lower costs, and mitigate the environmental effects of its high dependence on fossil fuels. A fundamental policy decision is whether the three grids that presently cover the contiguous US (and much of Canada) should be maintained as separate entities, tightly connected into one fully integrated continental grid, or broken into smaller separate modules to eliminate the possibility of geographically extensive blackouts.

The present three-grid structure, clearly recognizable by the late 1970s, was neither planned nor foreseen much earlier. There still is no federal agency administering a nationally integrated electric transmission network, as there is (in cooperation with the states) for the Interstate Highway System, and before that for the first east-west and north-south continent-spanning US highways built in the 1920s (Swift 2011). Yet there is no basis for concluding that the national power system grew irrationally or chaotically, like the internet. Indeed, there is no point in speaking of a "national system" before World War II because electrification was a local phenomenon, occurring simultaneously at many points across the map. For profit-seeking businessmen, engineers, and regulators, planning and expansion were highly rational in their local scope of operation, aimed at bringing electricity to more customers who were not too great a distance away, in higher amounts, at lower rates, and with greater reliability. A few, like Gifford Pinchot, first chief of the US Forest Service and governor of Pennsylvania, envisioned "giant" power grids covering a state or two, and there was the practical development of the Tennessee Valley project begun during the Depression, but these were not regarded as first steps for continental networks such as had developed for railroads and telegraphic communication.

Historic maps of very high-voltage transmission lines show that the grids, as we know them today, expanded rationally from five nuclei that were initially built in the 1920s and 1930s. Four of them (in California, Pennsylvania, Washington, and New Hampshire) delivered cheap hydropower, and one in Illinois delivered cheap coal-fueled electricity, to distant (for the day) urban loads. The demands of World War II stimulated more and farther expansion from these nuclei. Building continued more or less through the post-war boom until the 1990s, when problems besetting the industry and restructuring shifted the best investment opportunities from transmission to generation and marketing.

This process could have produced a single synchronous transmission network but for two barriers, one political, the other technical. The electric utilities of Texas, which found advantage to pooling their resources during World War II, did not want to be regulated by the federal government for interstate commerce. For that reason, they kept their operations within the state borders, even seeking court and legislative decisions to exclude AC/DC/AC junctions with the Eastern Interconnection as counting for interstate trade. The separation of the Eastern and Western Interconnections was due primarily to the few major generators and loads across the Great Basin and Great Plains, making it technically difficult to maintain dispersed elements in synchrony. Technology now available allows very high-voltage DC lines to span those distances efficiently. Other considerations must determine whether it is wise to do so. Comparison of the presently operating three grids, which differ considerably in size, gives no evidence that enlarging the continental grid will by itself improve efficiency or reliability.

PART IV

Energy controversies

10

RATIONALITY, PRO AND CON

Decisions about energy projects and policies are fraught with controversy. A few examples are the disputes over climate change, hydraulic fracturing ("fracking"), or permanent storage of nuclear waste. Sometimes people on one side think those on the other side are not simply wrong but arrogant, foolish, irrational, reckless, or crazy. Since I am emphasizing the merger (but sometimes opposition) between engineering and sociological perspectives, a good point of departure is the differing ways these disciplines view rational decision-making. By "rational" I mean based on reason, logical consistency, and a realistic appraisal of relevant facts and constraints.

Rationality is necessary for engineering. Applications of technology always occur within constraints of budget, schedule, physical limitations, and regulatory standards. Trade-offs are essential, perhaps reducing the size of a power plant to save money, choosing this design rather than that one to meet a safety standard or a completion date, opting for one fuel instead of another to comply with antipollution regulations. The primary if not exclusive imperative is to proceed rationally so as to maximize profit. Lee Raymond, CEO of Exxon Corporation (later ExxonMobil) from 1993 to 2005, commented, "We don't run this company on emotions. We run it on science and principles," on "the relentless pursuit of efficiency." Another executive who served on Exxon's board of directors commented, "They're all engineers, mostly white males, mostly from the South . . . They shared a belief in the One Right Answer, that you would solve the equation and that would be the answer, and it didn't need to be debated" (both quoted in Coll 2012: 55, 101).

Economics is the only social science that shares this imperative for rationality. It assumes that individuals and firms, faced with choices, deliberatively select the one that maximizes profit or net gain. The other social sciences do not enthrone rationality, giving considerable attention to other desiderata for decision-making including custom and habit, norms of equity, group pressure toward conformity, social structural demands, and what we may broadly call "emotional motives" for making one choice over another. Cognitive psychologists have identified many *heuristics*, simple guides for decision-making that help find adequate, though often imperfect, answers to difficult questions (Kahneman 2011; Slovic 2011; Fischhoff 2012). Often heuristics are used by the human mind in making choices quickly, jumping to conclusions that on slower reflection may or may not seem sensible. Anyone who has resolved to diet and then impulsively gorged on cookies and ice cream, or tried to quit smoking and failed, or has had sex in a manner that – or with a partner who – an hour later seemed a foolish choice, should understand that human actions and motives often are not grounded in consistent rationality.

There are other differences between the engineering and sociological professions. Engineers tend to be politically conservative, those in corporate employment more so than university professors of engineering. Social scientists (excepting economists) lean to the left politically, and they often favor populist interests over elites, citizen groups over corporations, the disadvantaged over the privileged (Rothman, et al. 2011). These biases promote different interpretations of the same events. What seems rational and therefore correct to the engineer may seem wrong headed to the sociologist, and vice versa.

For example, after the Exxon Valdez oil spill in 1989, sociologist Lee Clarke, studying the emergency recovery plan prepared by the consortium of oil companies that shipped oil out of Valdez, Alaska, found it rational on paper but impotent in practice. The plan was prepared to satisfy the State of Alaska's requirement that a spill response plan be filed. Perhaps thinking a serious spill would never happen, the response planners' goal was to satisfy the filing requirement, not to actually contain a spill. When a serious spill did occur, at night but otherwise under ideal weather conditions for an emergency response, the recovery plan was useless. Chemical dispersant could not be sprayed on the oil slick because there was no prior state approval, and in any case, the effects of dispersant on the oil and wildlife were uncertain. Floating booms intended to contain oil on the surface were unavailable, as were emergency boats supposed to be first responders (Clarke 2001). In this particular case, the planners' rationality satisfied the bureaucratic demand for an emergency response plan, but sociologically the plan was a fantasy, impractical in operation.

There is no free lunch

Engineers are aware that every source of energy has costs and benefits that must be weighed in planning a project. Opponents of a project often ignore its advantages, stressing only its liabilities, which seems emotionally irrational to the engineer who asks in frustration, "Where do you think your energy comes from?" For example, engineers tend to favor nuclear power because it can provide plentiful electricity without releasing any CO_2 or air pollution, advantages discounted by many opponents who concentrate on the dangers of radiation.

Disadvantages and benefits may change over time. CO_2 was not regarded as a problem until the late 1980s when global warming first emerged as a public concern. The comparative abundance and cost of fuels are prone to change with changing technology and political circumstances. In North America at the beginning of this century, natural gas was expensive and thought to be relatively limited in supply. Now it is plentiful and cheap, largely because of the new (and controversial) techniques of horizontal drilling and hydraulic fracturing, which are releasing a huge amount of methane that was recently thought to be inaccessible because it was "locked" in deep shale strata.

What we count as a "major" fuel is also in flux. Coal and oil were the primary fuels of the first half of the twentieth century. Natural gas and nuclear power came on the scene after World War II. We may hope that by the middle of this century, coal will be off the list, petroleum diminished in importance, and sunshine will hold a prominent place.

Rational planning

US Supreme Court Justice Stephen Breyer describes an egregious imbalance of costs and benefits in his argument for rational environmental decision-making. When he was an Appeals Court judge in 1990, a case that came to him arose out of a ten-year effort (with a 40,000-page record) to force continued cleanup of a toxic waste dump that had already been mostly cleaned. All the private parties involved had reached a settlement except one, who litigated the cost of about $9 million to remove a small remaining amount of highly diluted PCBs and volatile organic chemicals by incinerating the dirt. According to Breyer, all parties agreed that, without the extra expenditure, the waste dump was clean enough for children playing on the site to eat small amounts of dirt daily for 70 days each year without significant harm. Incinerating the soil would have made it clean enough for the children to eat dirt for 245 days per year;

> But there were no dirt-eating children playing in the area, for it was a swamp. Nor were dirt-eating children likely to appear there, for future building seemed unlikely. The parties also agreed that at least half of the volatile organic chemicals would likely evaporate by the year 2000.
>
> (Breyer 1993: 12)

What was the point of spending $9 million to protect non-existent dirt-eating children? Who beside the plaintiff would justify this irrational option?

An energy company must seek some reasonable balance between cost and benefit in choosing between two technologies or fuels, both capable of producing a given amount of energy but each having different advantages and disadvantages. How is this approached rationally?

First, such decisions are rarely without precedent, nearly always occurring within a set of organizations and institutions that have prior experience with one or more of the technologies or fuels. The first tendency of planners, not in itself necessarily rational, is to choose the technology with which they are experienced, or the one that other organizations in their set, in similar circumstances, have chosen – so long as the earlier choice worked out satisfactorily. In the 1960s, electrical utility companies in the US, prodded by the federal government's "Atoms for Peace" program and further encouraged by inducements from nuclear reactor manufacturers, began to build nuclear power plants. Having built one, they often built another, and then a neighboring utility company might follow suit. This two-decade trend ended partly because of public opposition to nuclear power after the Three Mile Island accident in 1979.

The second criterion is to maximize profit. This usually involves minimizing costs (ideally without sacrificing reliability or safety), counting both the cost of constructing new operating equipment and the cost of fuel consumed during the lifetime of operation, and accounting for government subsidies intended to boost a technology, as was the case with nuclear power and is presently important for adoption of renewable energy sources. If costs are minimized sufficiently, some of the savings can be passed on to consumers, who increase their consumption of the cheaper energy, increasing the total profit to the company. The cessation of nuclear power plant construction in the 1970s was essentially caused by the hugely (unanticipated) escalating cost of bringing a new nuclear plant on line, which became on the order of $5 billion, more so than by public opposition. Also in the 1970s, petroleum prices skyrocketed, so utility companies stopped using oil as a fuel for electrical generators, turning to far cheaper coal, which today is the primary source behind nearly half of America's electricity and is still profitable though increasingly pressed by environmental regulations. Most recently, new electrical generators tend to use gas turbines, which are quick and inexpensive to install, and are fueled by relatively cheap natural gas.

During the first half of the twentieth century, these two considerations – prior experience and profit maximization – explained much of the behavior of energy companies. With the rise of the modern environmental movement in the 1960s and 1970s, public concern over blatant pollution pushed the US Congress to create strong environmental legislation limiting air emissions and water effluents, and protecting endangered species. These laws required large new projects that depended on federal funding to submit environmental impact statements that evaluated potentially negative impacts of the projects. The other industrial nations soon passed comparable legislation. (Some Third World nations, notably China, passed similar laws, but with a large gap between regulations on the books and actual compliance and enforcement.) These new requirements addressed what economists call "externalities," or nonmonetary costs such as pollution and habitat destruction that did not figure directly into a corporation's profit accounting. They could have been incorporated into the accounting if they were given a monetary value, say, by charging a federal tax on air or water pollution, but this was not done. Instead the new environmental protections usually took the form of mandated limits on pollution, or requirements to use the most non-polluting equipment available, operating as constraints on profit accounting rather than being integral to it.

As a result of these reforms, we can speak of three kinds of "costs" explicitly incurred in the production of energy. First is monetary expense, explicit in the profitmaking of corporations. Second is environmental damage, not unambiguously expressible in monetary (or other quantitative) units but nonetheless turned into explicit burdens for energy producers (and indirectly for their customers). The third cost that became explicit during this period, *and like monetary cost was made countable*, is the health risk to workers and the public, to which we now turn.

Risk-benefit analysis

How many lives are lost by getting a given amount of electricity from nuclear power compared to getting it from coal or the sun? Knowing this, we can make a rational decision to minimize the number of lives lost in obtaining our energy, if that is our goal.

The roots of risk-benefit analysis go back centuries to when insurers calculated the odds that a sailing ship on a years-long trading voyage would return safely with its valuable cargo intact. To avoid losing all the money he had invested, a ship owner (who stayed home) laid a bet against the ship's return by paying an amount of money – a premium – to the insurers; if the ship foundered the insurers would pay the owner the value of whatever was lost. The insurers, on the other side of this transaction, effectively bet that the ship would return safely, thus keeping the premium without paying for any damages. It was a mutually beneficial arrangement. Insurers who calculated the odds properly made a good living, despite having to pay out occasionally. Ship owners were not ruined by an occasional loss because in that event they were compensated by the insurers.

Modern risk analysis came to the fore in the 1970s, at the height of the public controversy over nuclear power, as an attempt by nuclear industry spokespeople to demonstrate quantitatively that the health risk of nuclear power was small compared to the health damage from fossil fuels, even allowing for occasional nuclear accidents (Starr 1969; Mazur 1985). Risk analysis now permeates technology development, business decisions, and environmental regulation. Here is a simple illustration, based on original work by Herbert Inhaber as modified to incorporate later critiques (Inhaber 1979; Fritzche 1989; Mazur 2007; also see Burtraw et al. 2012).

If we considered only the operation of a generating facility already in place, then obviously coal with its air pollution or nuclear power with its potential for accidents are riskier than a photovoltaic system, which simply collects sunshine. But we must consider the entire generation cycle, including material and fuel production, component fabrication, plant construction, operation and maintenance, transportation, and waste disposition. Remember that solar and wind are low-density energy sources, so a very large array of collectors is needed to obtain the same total energy that is produced from nuclear or fossil fuels in more compact facilities. To build the photovoltaic facility in the first place required a lot of material, including aluminum, which itself requires a lot of electricity to refine. Occasional accidents from this material acquisition, transportation, and construction must be counted into the overall risk of the facility. Solar and wind do have appreciable risks when the entire fuel cycle is included, more so than is usually recognized. Most of these costs must be borne before any energy is generated; once solar and wind collectors are in place, they are relatively risk free until they must be serviced or replaced. Also, since solar and wind-powered generators work only when sunshine or wind is available, these systems must include some backup source of power that will supply electricity at night or on calm days, and so their risks must be counted too. We will assume that the backup source is natural gas, a relatively safe fuel that does not add much to the overall risk of solar or wind.

Hydropower and nuclear power have especially high potentials for catastrophic accidents: dam failures or major releases of radiation. How, then, does one compare Technology A, which kills one person yearly, with Technology B, which usually kills no one but has a one-in-a-thousand (0.001) probability of killing a thousand people in a single accident? Risk analysts handle this by calculating the "expected" number of yearly deaths from rare accidents, defined as the probability that the accident will occur in any year multiplied by the estimated number of deaths if it did occur. For Technology B, that would be 0.001 x 1,000 = 1, which is the same as the routine yearly death from Technology A. In effect, the deaths from a rare accident are averaged out over all the years of operation.

The calculated risk associated with each technology is available elsewhere (Mazur 2007: 188), but since these numbers are imprecise, I emphasize only the general pattern of results. Overall, coal and oil are far riskier to health than several comparably low-risk options: nuclear power, natural gas, hydroelectricity, wind, and solar power.

Risk analysis shows nuclear power to be surprisingly safe in this comparison, which even counts risks of radioactive waste disposal. (This is why the nuclear power industry was among the earliest proponents of risk-benefit analysis.) The largest part of nuclear power risk comes in construction and fabrication of fuel. During operation, there is no air pollution. Although the potential for a catastrophic accident is present, its estimated probability is very low, so the yearly expected risk – averaged out over the years of operation – is low. If we take the (imprecise) risk calculations literally, nuclear power is one of the safest energy sources, while natural gas is safest of all.

In interpreting these comparisons, it is important to recognize significant risk factors that have been left out of the calculations. The calculation of nuclear power's risk takes no account of the possibility that radioactive material from reactors can, under certain conditions, be converted into weapons. The calculated risk of oil does not include wars fought to protect foreign sources of petroleum, as in Iraq, nor the funding of terrorist activities from the huge oil revenues paid to oil-producing nations in the Middle East.

A critique of risk-benefit analysis

Economists conceptualize actors – people or firms – as utility maximizers. When confronted with a choice among options, these actors attempt to select the one that gives them the greatest net satisfaction, or "utility." Analogously, the best political situation for a group is postulated to be the decision that maximizes the group's utility. Of course, the postulation of utility maximization is a tautology unless we have some way of operationalizing "utility," for otherwise it is simply "that which is maximized." Nonetheless, this is a convenient starting point, for it allows us to compare the risk-benefit approach to decision-making with other, equally defensible methods of making decisions. Too often risk analysts regard their particular rationalist perspective as superior to others, when, arguably, it simply reflects their own parochial viewpoint.

The risk-benefit perspective assumes that each policy option has some good features and some bad features. Good features might include financial profit, personal happiness, and environmental beautification; bad features might include financial cost, resource depletion, pollution, and, of course, risk to life and limb. A crucial assumption is that one can *and should* balance the positives against the negatives, comparing the weights of one against the other in order to find that option that gives the greatest net profit of goods over bads. The idea that good features of an option can offset (or be "traded off" against) bad features is central to the risk-benefit perspective, as is the notion that doing nothing, i.e., avoiding a decision, is itself a choice that has good and bad consequences.

Under certain scope conditions, the risk-benefit perspective is clearly superior to other means of arriving at a policy decision. Principal among these scope conditions are: (a) that goods and bads are *commensurable* in the sense that they can be measured by some common metric so that one can be compared against the other; (b) that goods and bads *should* be traded off against one another; and (c) that the person or group can maximize its utility by maximizing its net profit of goods-minus-bads. It is essential for the risk assessor to realize that these scope conditions do not apply to all policy decisions. Furthermore, in the case of controversial policies, or highly emotional ones, one side may be willing to accept these scope conditions, but the other side may not. In such cases, communication across sides is doomed to failure, unless each side respects (or, at least, understands) the scope conditions that the other (often implicitly) accepts or rejects.

In highly emotional cases, especially those involving political controversy or strong ideology, it is unlikely that both sides will accept the scope assumptions that goods *should* be traded off against bads. The abortion controversy is an obvious example of this. The pro-abortion ("pro-choice") side does accept this scope assumption, arguing that the bad features of abortion are offset by the good of allowing women their own choice, of improved family planning, of population limitation, etc. But the antiabortion ("pro-life") side argues that *no* amount of benefits can offset the heinous sin of murdering an innocent unborn child; thus they reject the assumption that what is good should be traded off against what is bad.

Sociologist Charles Perrow (1999) similarly rejects this scope assumption in asserting of nuclear power and nuclear weapons that there is no way their inevitable risks can be offset by any reasonable benefits. He spoke for many opponents of nuclear power plants who argued that the risk of a catastrophic radiation release – even though improbable – cannot be justified by the advantages of nuclear-generated electricity, nor even by the higher "routine body count" produced by other methods of generating electricity that do not entail the potential for catastrophe.

While many people disagreed with Perrow about nuclear power plants, during the Cold War hardly anyone disagreed with him about nuclear weapons. By almost anyone's calculation, the risk of a nuclear war between the United States and the USSR, each armed with about 10,000 warheads, exceeded any conceivable benefit, and yet both nations continued to expand their nuclear weaponry, suggesting that neither country accepted the scope conditions of risk-benefit analysis in planning its defense. This should not be surprising, for there is a long tradition regarding nationalistic goals as sacrosanct, and considering no sacrifice too great in achieving them. When Teddy Roosevelt proclaimed, "Millions for defense, but not one cent for tribute," he was not speaking from the perspective of a risk assessor.

Like Teddy Roosevelt, or the Pope in his denouncement of abortion, or the parents who will pay any price for their own child's safety, or the antinuclear activist with extreme fears of a radiation catastrophe, we all identify certain goods as so precious and certain bads as so abominable that we insist they should not be traded for any compensating measures. In this instance, we reject the perspective of risk assessment, that goods should be balanced against bads. To the contrary, we argue from the perspective of *inviolate principle* that no compromise should be permitted. We will maximize our utility only by defending the greater good and perhaps going down fighting, but certainly not by compromising our principles or selling out. In such circumstances, risk analysts will waste their breath arguing their way is better.

The perspective of inviolate principle, which rejects the propriety of trade-offs, is not the only alternative to the risk-benefit perspective as a means of maximizing utility. If we reject even the possibility of trade-offs (thus rejecting the scope condition that goods and bads are commensurable), then we arrive at a position that may be called the *humanist perspective*. This position is so well known that little discussion is needed here. In essence, it says that one cannot appraise qualities as diverse as clean air, flourishing rainforests, preservation of species, human life, and industrial profit in any common currency. To even attempt such comparisons degrades the human condition.

Risk analysts have probably given more attention to this criticism of their position than any other, especially in the many discussions of the dollar value of a human life, without providing a satisfactory response. They seem particularly vulnerable in using expected values to represent risk, equating a known and recurring number of deaths with the probability-times-consequence of an unlikely – but very costly – accident. One hundred widely scattered deaths per year simply is *not* the same as a one-in-ten chance of a catastrophic loss of a 1,000-person community.

The humanist perspective recognizes that such mathematical juggling can be substantively meaningless, asserting that each major policy decision – whether involving trade-offs or not – is unique and must be decided on its own merits with full appreciation of its historical and cultural context. Imagine how much weight risk-benefit calculations will carry in Japan the next time someone proposes to build a new nuclear power plant.

One other perspective deserves mention here, that of *political pluralism*. In effect, it rejects the scope condition that a society's maximum utility can (and will) be achieved by maximizing the net profit of goods-minus-bads. The pluralist emphasizes that society is made up of diverse groups each with its own interests and each defining what is good and what is bad in its own way, which is often inconsistent with the views of the other groups. Society cannot maximize its utility if the groups cannot agree on what is good and what is bad. Thus the pluralist sees little to be gained from the risk-benefit perspective, except, perhaps, in those rare instance of national consensus. For the pluralist, the preferred method of decision-making is through the usual political process, which no doubt fails to maximize any hypothetical tally of goods-minus-bads, but which does offer opportunities for negotiation and compromise so that each group can bargain to enhance its own wellbeing.

The difference between the pluralist and the risk-benefit perspectives is most obvious when we consider the problem of equitable distribution of society's goods and bads among its various groups. The risk perspective is concerned solely with the amounts of goods and bads in the society. Often one group (say, industrialists) receives most of the goods, while another group (say, workers or the poor) suffers from most of the bads. A simple redistribution from a state of unjust inequality to one of more equity would not increase goods-minus-bads, yet, from the pluralist perspective, it would raise the utility of the society. Some pluralists are so wary of the risk-benefit perspective – with its willingness to balance one group's risks against another group's benefits – that they regard it as actually inimical to the wellbeing of a fair-minded society.

Conclusion

There are good arguments for the engineering/economics worldview of rationally balancing costs against benefits in a firm's decision-making and in governmental regulation (Revesz & Livermore 2008). For corporations, this has always translated into maximizing profits, and that remains their primary goal. However, in the past half century, at least in the industrial nations, important new factors must be considered that heretofore had been regarded as external to the business of supplying energy. Some of these factors cannot be given unambiguous monetary value, such as environmental protection or species preservation; these have been added to the accounting in the form of regulatory requirements, standards, or constraints, or at least requiring an environmental impact statement that evaluates potentially adverse impacts of a proposed project.

If early industrial titans did not always show much concern for the risk to health of their workers and the public, this is certainly a requirement today in the developed nations, usually implemented by environmental health and safety regulations. It is impossible to achieve absolute safety. A rational perspective presumes a tradeoff between health risks, on the one hand, and costs of protection on the other. In the 1970s this perspective was pursued vigorously by the nuclear power industry, trying to convince a skeptical public that, overall, nuclear reactors were safer for producing electricity than coal- or oil-fueled generators. Despite the more or less valid (but limited) conclusions of this approach, it did not succeed in winning a positive consensus for nuclear power, especially after the accidents at Three Mile Island, Chernobyl, and most recently at Japan's Fukushima Daiichi nuclear complex. Nuclear opponents did not trust the industry's numbers, and more fundamentally, rejected the rationalist approach with its embedded biases.

Risk-benefit analysis is friendlier than other perspectives to options that have a small but real possibility of causing a catastrophe, for it counts the catastrophic death of a whole community no differently from the routine background of scattered deaths. It is sometimes inappropriate because it recognizes no goods as so precious and no bads as so despicable that they should not be traded. Since everything is negotiable, the perspective is more felicitous to those who accept compromise than to those committed to inviolate principles. It is friendly toward options that channel goods to one group and bads to another, for risk-benefit analysis is concerned only with the goods-minus-bads over the entire society, and not with the fairness of their distribution. Recognizing these biases, risk analysts should not be surprised when opposing parties, coming from different but equally justifiable perspectives, fail to recognize the cogency of their conclusions.

11

THE DYNAMICS OF TECHNICAL CONTROVERSY

Benjamin Franklin showed that lightning is electricity with his famous kite-flying experiment and soon gave his discovery practical application by inventing the lightning rod. It had been the custom for centuries in Europe to dispel lightning from thunderstorms by ringing church bells. The tenacity of this practice was remarkable since high church steeples make excellent targets for lightning. Furthermore, electricity easily travels down a rain-wetted bell rope to the ringer. One eighteenth-century European source reports that over a 33-year period, lightning struck 386 church steeples, killing 103 bell ringers.

Some clerics immediately accepted Franklin's rods as an improvement over bell ringing, but others in the church as well as some laypeople opposed the innovation for a variety of reasons. There was the religious objection that if God wanted to strike a building, it was presumptuous of man to interfere. One leading "electrician" of the time, who had earlier attacked Franklin's theory of electricity as contrary to his own, claimed that the rods were more likely to attract lightning to a building than to preserve the building from a strike. The controversy was particularly heated in Boston where the Reverend Thomas Price warned that electricity, transferred from clouds to earth via metal rods, would concentrate in the ground. The city, where many rods were installed, would accumulate a local pool of electricity that could enhance the likelihood of earthquakes (Cohen 1952).

Technical disputes that become public controversies

Public controversies over science and technology have been around a long time and for decades have been studied by social scientists (e.g., Nelkin 1979; Mazur 1981; Clarke 1991; Kasperson & Kasperson 2005). While each controversy is unique in some ways, nearly all follow common lines: once in the public arena, staunchly opposed parties dig in their heels and rarely concede any point in dispute. They rarely engage their differences face-to-face, point by point, but make their cases obliquely to third parties, often through mass media, public meetings, or on the internet, often arguing past one another, and with little check on the validity of their statements. Proponents of a technology emphasize its benefits and minimize its risks; opponents do the opposite. They quote data or theory that supports their position and discredit or reinterpret studies that go against them. If one side comes close to nailing down the falsity of a claim by the other, the other side will raise a new claim, analogous to a game of Whac-A-Mole. Scientific or technical disagreements that enter the public arena are correlated with differences in politics, ideology, or economic interest.

Virtually never are these disputes resolved in the sense that adversaries come into agreement. But inevitably they are "socially resolved" by wide acceptance of some practice or policy (e.g., installing lightning rods) or simply through loss of interest by the media and the public.

The case of the lightning rod is misleading if it suggests to the reader that opponents of a technology are usually wrong. A survey of 31 technology-related health warnings appearing in the news media of the 1950s and 1960s concluded in retrospect that about half were correct (Mazur 2004). Technical controversies can be highly obstructionist, on the one hand, but they can also serve as a brake on wanton development and, when occasionally properly channeled, as a useful form of technology assessment.

There are important ways in which technical controversies today differ from those of Franklin's time, or of the "battle of the currents" between Edison and Westinghouse. The established news media (including their internet outlets) now play a major role, not so much by taking sides, because their coverage of controversial issues is normally balanced, but by giving high coverage to the dispute and thus giving it legs as a public concern. Private parties and interest groups using the internet or social media play a similar role in publicizing a controversy, but here advocacy for one side is common, and while much information conveyed via the web is solid, much is erroneous or purposively inaccurate. Another difference is that scientific and technical arguments in modern disputes have become so esoteric that only specialists can truly assess conflicting factual claims, so most of us don't even try. Usually we opt for the side in these disputes that is congruent with our political or ideological views, not because we have any clear understanding that one side is factually correct or wiser.

At the time of writing there are numerous controversies in the news over one aspect or another of energy policy, among them climate change; hydraulic fracturing ("fracking"); opening new areas for oil extraction, either offshore or in the frozen North; risk of accidents from deep-sea oil drilling rigs (following the *Deepwater Horizon* spill) or nuclear power plants (following Japan's Fukushima disaster); limiting very-small particulate emissions from coal-fired power plants; and the siting or routing of new oil or gas pipelines, high-voltage electrical transmission lines, and fossil-fueled, nuclear, or renewable energy facilities. All of these combine factual disputes (i.e., adversaries disagree about whether "objective" facts are *scientifically true or false*) with value disputes (i.e., adversaries disagree about what *should or should not* be done). In studying such controversies for years, I nearly always find a high correlation between adversaries' positions on the facts and their positions on values or their economic interests. I focus this chapter on the dispute over climate change, which is bigger and has more long-term import than most others, but in terms of general dynamics is unexceptional as a technical controversy.

It is worth pointing out here the truism of political philosophy that facts alone cannot determine social policy; value judgments are necessarily involved in deciding what actions are right or wrong. It may be unwise, but it is not logically contradictory to accept as scientific fact that human activity is warming the atmosphere and oppose any government policy to slow this trend. Perhaps alleviative actions seem too expensive for the nation or for one's own business; one might live in a region that would benefit from warming; one might be opposed to government regulation as a general principle; one might believe that climate change is God's prelude to the Rapture, and so on. If not logically necessary, it is nonetheless empirically true that belief in anthropocentric global warming and belief that government action should limit greenhouse gases tend to occur together, while opposition to government action is often associated with disbelief in the science of global warming.

Disputes between experts

Any case study of a technical controversy will contain references to each side's experts – the properly credentialed scientists, engineers, and physicians who buttress its positions with technical information and who undermine the scientific basis of the other side. Laypeople are often confused and dismayed when one scientist contradicts another's facts. There is a popular conception that scientists know the truth – at least in their domains of inquiry – and if two of them disagree, then one must have lied or made a mistake. Of course, science is not equivalent to truth, and there are ample opportunities for two competent scientists to disagree with one another, particularly on issues at the state of the art. The process of science – the requirement that theory be consistent with empirical observation, and that scientific publication must be submitted to peer review so their methods and conclusions pass muster before expert evaluators – is not infallible but operates well as a self-correcting system of knowledge.

From a *purely scientific* point of view, there is no doubt that the global climate is warming, very likely because humans are massively burning fossil fuels. If politicians, policymakers, and the public were as rational as Mr. Spock, there would be no publically aired dispute about these factual matters. But Mr. Spock lives only in television reruns and movies. From a sociological point of view, the facts are only a part – perhaps a small part – of the larger political controversy in which they are embedded. That a few scientists deny these conclusions of climate science (and related sciences) is irrelevant because one can fairly easily find dissenting experts for any topic about which there are politically or legally competing constituencies.

There are bona fide scientists who insist that the fluoridation of drinking water causes cancer, that cold fusion is a real phenomenon, that AIDS is not caused by HIV infection, that aliens have visited Earth, or that the Bible is literally true and Noah's flood was a historic worldwide catastrophe (Mazur 2008).

Many of the scientific issues surrounding climate change are highly complex, but a few of the early factual ambiguities were sufficiently simple enough to be explained here. In 1988, when global warming first became a public issue, there were good scientific reasons to be skeptical that it was actually occurring, and if it were, that human actions were the cause. It was known that earth's climate in the long past has been far warmer and at other times far cooler than now, so a present warming, if true, might be due to natural causes. The veracity of average thermometer readings over the past century was then uncertain. Even accepting the published century-long temperature trend, it had a puzzling feature that seemed to contradict the assertion that increasing CO_2 was heating the planet. Figure 11.1 shows that from about 1940 to 1965, average temperature was *decreasing* while CO_2 (extrapolating from the Keeling Curve) was steadily increasing. (The "temperature anomaly" plotted on the graph is the difference between a year's average temperature and the mean global temperature for the period 1951 to 1980.) If rising CO_2 causes rising temperatures, why at that time were the two trends moving in opposite directions? Also, there were inadequately evaluated feedbacks that might either mitigate or accelerate warming trends. For example, an expected increase in atmospheric water vapor might intensify the heating, since H_2O is a

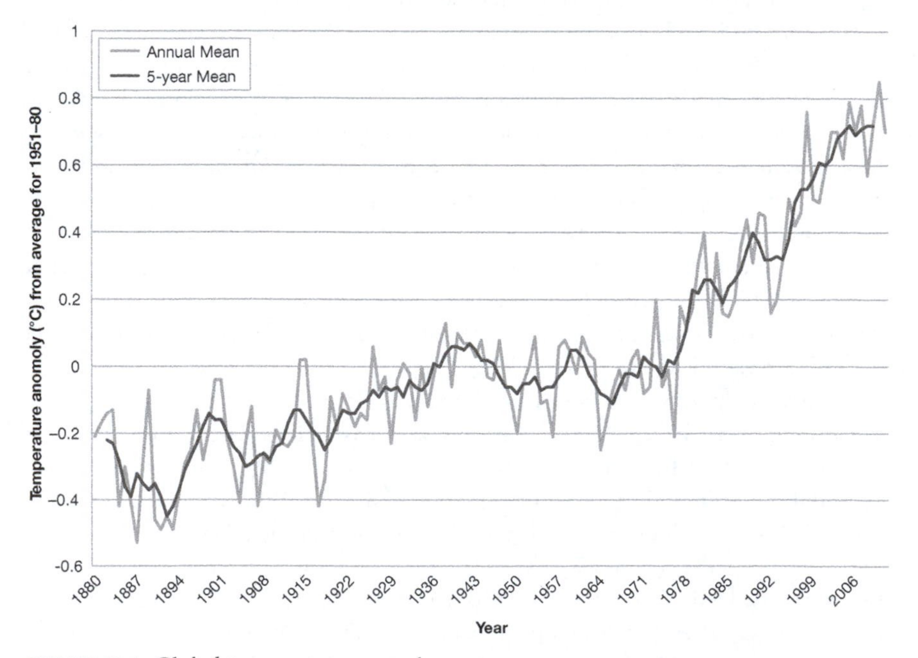

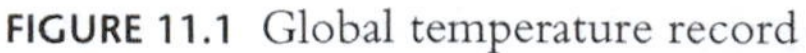
FIGURE 11.1 Global temperature record

greenhouse gas, or it might lessen it, because increased cloudiness would cause more incoming sunlight to reflect off the earth's atmosphere.

With the evidence ambiguous, there was ample reason for competent, earnest scientists to doubt that human activity was warming the climate, and to argue against those making such claims. Furthermore, no one knew all the consequences of a warming climate, should it occur. Some plausible effects could be beneficial for certain regions, such as longer growing seasons at high latitudes in North America and Eurasia. Higher concentrations of CO_2 spur plant growth, likely increasing agricultural yields. Melting polar ice above Canada would open a Northwest Passage, long but futilely sought by early explorers as a northern sea route between the Atlantic and Pacific Oceans.

Attempting to resolve the science of global warming

Almost never do opposing experts in an emotionally polarized public controversy come to agreement over their factual disputes, with one saying, "I was wrong, you were right." (But see Muller [2012] for the unusual reversal by a climate skeptic.) How, then, can policymakers and the public incorporate "correct" science and engineering into their decision-making? Indeed, one might wonder how policy *ever* gets made in a technologically complex society.

In fact, myriad technology-laden policies are made yearly without a peep of public controversy or any notice in the news media. Government agencies and professional engineering societies routinely provide or upgrade standards for construction of buildings to withstand earthquakes, fires, or floods; safety and compatibility codes for electricity and water supplies; standards for highway, shipping, and airline operation and safety; dams and bridges; pollution controls; use of the electromagnetic spectrum for communication; food safety; and so on. Overall, this process works well, though some of these standards later prove short-sighted. Structural designers of the World Trade Center, knowing that an airplane once hit the Empire State Building, planned the twin towers to withstand such an event but in retrospect inadequately because both towers collapsed in exactly the same way roughly an hour after the impacts. This was sufficient time for nearly everyone working under the impacted floors to escape, while those above died (Wright 2007). Had the hijackers anticipated this, they likely would have aimed for lower floors. Too bad there was no public controversy over the design of the towers, which might have better aired plausible risks and their mitigation.

Very few projects or processes become the focus of public technical controversy. In the few that do, when polarized disputes become vivid under the spotlight of media coverage, routine methods of policymaking do not work. Sometimes especially problematic disputes are referred to a prestigious scientific commission or organization for resolution. In the United States it is often the National Academy of Sciences (NAS), established by Abraham Lincoln to honor esteemed scientists but also to provide scientific advice to government when needed. Yet even reports of prestigious commissions or NAS committees, which may not be unanimous, seem impotent to settle raucous politicized controversies (Boffey 1975).

The Intergovernmental Panel on Climate Change (IPCC) is the first major international organization with a similar mission, established in 1988 by the United Nations and the World Meteorological Organization to assess the evolving scientific understanding of climate change, its impacts, and the potential for mitigation (http://www.ipcc.ch/). While the IPCC can assess the probable results of various policy options, it cannot recommend specific actions. The intent was to use scientists to make scientific assessments, not to propose government policy.

Issuing its first assessment in 1990, the IPCC releases updated reports about twice per decade. These assessments are extraordinarily comprehensive and transparent. For each report there is first a round of expert review, including comments from thousands of scientists. Lead authors for each chapter must consider and respond to all comments and make appropriate revisions. The revised draft is reviewed again, this time by representatives of all participating United Nations member nations as well as climate experts. The final draft of the all-important summary for policy makers is agreed upon, word for word, in a plenary meeting of government delegations. At the 1995 plenary in Madrid, there was an intense debate between scientists authoring the report and the delegate from Saudi Arabia who said the word *appreciable* was too strong in the proposed summary statement: "The balance of evidence suggests that there is an appreciable human influence on climate." After two days of argument, a compromised was reached by replacing *appreciable* with *discernible*, and that is how the summary was published (Mann 2012).

As the pace of research increased, results became clearer. Temperature trends were verified, and proxy analyses of global climate extended backward in time. It was recognized that industrial activity during World War II and the post-war period added sulfate aerosols to the atmosphere, reflecting incoming sunlight away from the earth; hence the cooling trend of those years. In 1991, when the Mt. Pinatubo volcano in the Philippines spewed sulfate aerosols high into the atmosphere, there was a similar (and predicted) abatement in the warming trend, lasting until the sulfates dispersed and warming resumed. (Ironically, successful attempts to remove sulfur pollution from the atmosphere also removed this cooling effect.) While effects of water vapor and other feedbacks remain uncertain, the case was increasingly solidified that warming was at least partly due to human burning of fossil fuels. Among atmospheric scientists, excepting very few contrarians, this is nearly consensually accepted today.

Higher global temperatures would surely raise sea level, a result of two well-known effects. First, glacial ice *that now rests on land* would melt and flow into the ocean basins, increasing the amount of sea water. Greenland and Antarctica together hold enough ice that if wholly melted – an unlikely prospect – would raise sea level many meters. (Ice floating on the sea as icebergs or the Arctic ice at the North Pole would not raise sea level by melting because Archimedes' Principle tells us that two floating objects of the same weight, whether solid or liquid, displace the same volume of water. You may test this by marking the level of water in a glass containing floating ice cubes, and then see if the level changes when the cubes melt.) The second cause of rising sea level is that water (above 4° C) expands when warmed, so as temperature rises there will be an increase in the volume of sea water, apart from any new melt water flowing in from the land. Unless there is abatement of the greenhouse gases now entering the atmosphere, we will see higher sea levels by the end of this century, causing loss of costal and low lying lands around the world. Hurricanes would likely become more powerful, fueled by warmer ocean water, and there would be an increase in other extreme weather events including floods and droughts, changes of natural and agricultural habitats, and changes in the geographic distribution of insects, animals, and humans.

The summary of the most recent IPCC report, the fourth assessment, released in 2009, reads in part:

> Warming of the climate system is unequivocal, as is now evident from observations of increases in global average air and ocean temperatures, widespread melting of snow and ice and rising global average sea level . . . Most of the observed increase in global average temperatures since the mid-20th century is *very likely* due to the observed increase in anthropogenic GHG [greenhouse gas] concentrations.
>
> (Pachauri & Reisinger 2009)

Perhaps naively, the IPCC founders thought they could keep the objective science of climate change above the fray, leaving the battling over what to do about it to politicians and lobbyists. They were wrong. Contrarians seized on a single paragraph in a 938-page volume that overstated the rate of recession and date of disappearance of Himalayan glaciers. News media played it up. The IPCC later acknowledged that these estimates were poorly substantiated.

This nearly coincided with a hacking of the computer system at Britain's University of East Anglia by someone operating via an anonymous server located in Turkey, downloading personal emails among climate scientists that, read selectively, inferred shenanigans about misrepresenting data and impeding activities of opposing scientists. Former vice presidential nominee Sarah Palin was among the first to criticize climate researchers after the hack. Journalists called the brouhaha "Climategate," contrarians called it a "smoking gun" that revealed the perfidy of climate researchers (Mann 2012). This was Whac-A-Mole tactics, shifting the focus of controversy from the substance of climate change to ethical violations and exaggerations by the IPCC and its affiliated scientists. Probably the reason it was highly covered in the news, and generated such ferocious political commentary, was its timing during the run-up to the December 2009 climate summit in Copenhagen. Institutional hearings later cleared the researchers whose emails had been hacked of improper behavior. The Copenhagen summit failed to produce an agreement on limiting greenhouse gases, though for more weighty reasons than this tempest in a teapot.

Partisans

Who are the people – scientists and nonscientists – who become involved in a technical controversy? How do they choose sides, and do they ever change sides? What motivates them?

To avoid a common point of misunderstanding, it is necessary to differentiate between active participants in a controversy and passive members of the wider public who only occasionally express their views, perhaps in response to a question on an opinion poll, or by posting a comment on Facebook, or spending a day listening to bluegrass music at a demonstration. The demographic and motivational characteristics of these two classes of partisans may be quite different, a point that is lost by some commentators.

The activists, few in number, often participate in the controversy as part of their occupational role or auxiliary to it. This includes people in trade and lobbying organizations, government and corporate officials, academic scientists and attorneys, employees of environmental organizations and policy institutes, journalists and filmmakers. Often their occupational position defines their side of the controversy: environmentalists favor limits on greenhouse gases; employees of energy companies are opposed. The common pattern is to support the policy of one's organization. Apart from any personal motivation, the major reason these activists participate is that it is part of their job, and their positions are likely to reflect the attitudes promoted among their colleagues, clients, sponsors, and professional associations.

Voluntary activists, though often important, are necessarily limited by the time and resources they have available apart from their obligations to jobs and family. They may have a personal stake in the outcome, perhaps seeing a threat to their home and family, or being concerned about nearby land use. Their participation is greatest when the controversy is "hot," that is, when it is receiving a lot of attention in the news media. Since media attention ebbs and flows, their participation correspondingly rises and dips, sometimes in waves that last over years. They rarely act alone, instead participating in a network of like-minded friends and associates. Local groups with shared interests make contact, coordinate activities, and often national leaders emerge, sometimes in a semi-professional role so they have the wherewithal for sustained activism. Usually they have a prior political or ideological orientation that predisposes them to one side or the other.

Polarization is a strong motivator. As disputes become heated and publicized, the sides become more sharply drawn, and individuals become committed to a position that before was only a casual identification. From the outset of the climate controversy, when the scientific case was ambiguous, two highly respected climate scientists, James Hansen of NASA and Richard Lindzen of MIT, took opposing positions from which they have never moved. At that time Hansen claimed without scientific justification that the severe US heat wave of summer 1988 was caused by global warming. Subsequent scientific analysis has greatly strengthened the case for human-caused climate change, but it is still indefensible to blame any single weather event on long-term climate change. Lindzen's objections to anthropocentric warming have changed over the years, most recently emphasizing scientifically unresolved questions about the mitigating effect of clouds (Stevens 1989, Gillis 2012).

Fossil fuel companies, and conservative political groups and think tanks, fund and promote studies that refute or discredit the science of global warming in the public mind, attempting to avoid government regulation that would diminish the use of fossil fuels. To put this in context, readers should understand that similar tactics occur in nearly any politically salient technical controversy, and on both sides. Like-minded scientists and activists, and allied support groups, coalesce to fortify their own position and undermine that of the other side with selective or exaggerated argumentation. That's the way these technical-political arguments work (Mazur 1981).

The climate disinformation campaign is unusually severe, comparable to that waged by tobacco companies to cast doubt on the science linking smoking to lung cancer (Oreskes & Conway 2011; Mann 2012; Proctor 2012). But we must set any scientific "guns for hire" against a Richard Lindzen, who is not funded by industrial or political sources, or the long-serving CEO of ExxonMobil, Lee Raymond, who as recently as his retirement in 2005 sincerely doubted that global warming was occurring (Coll 2012). Climate contrarians are not ipso facto venal or dishonest, but almost certainly they are wrong.

In the United States, among the public at large, the controversy over climate change has become deeply enmeshed in partisan politics. In 1998, when the Gallup Poll first asked Americans if they believed the seriousness of global warming is exaggerated in the news, there were no great differences in the perceptions of Democrats, Republicans, and Independents (Figure 11.2). Starting in 2001, the question was repeated yearly, showing increasing divergence during the first decade of the century. This was likely reinforced by Al Gore and his documentary, *An Inconvenient Truth*. Gore's warning about the climate, in sharp contrast to then-President George W. Bush's refusal to support the Kyoto Protocol to limit greenhouse gas emissions, may partly explain the polarization in views about climate change between Republicans/conservatives and Democrats/liberals. This is a blatant example of politics trumping science, but it is not unprecedented or even unusual. It is easy to forget that environmental protection began as a bipartisan issue, and that the most significant American legislation to protect the environment was signed by Republican President Richard M. Nixon.

Apart from party differences, Figure 11.2 shows unison year-to-year changes in belief that the news about global warming is exaggerated. This waxing and waning will be taken up in the next chapter, but it is relevant here in the context of rising and falling activism, especially among volunteer partisans. Typical of technical controversies and of protest movements generally, levels of concern, commitment, and activism fluctuate over time, often for reasons that are external to the substance of the controversy itself. We may visualize partisans as surfers riding successive waves that are larger national issues. As each wave diminishes, the technical controversy falls unless it can catch another wave. Each wave, to be suitable, must have clear

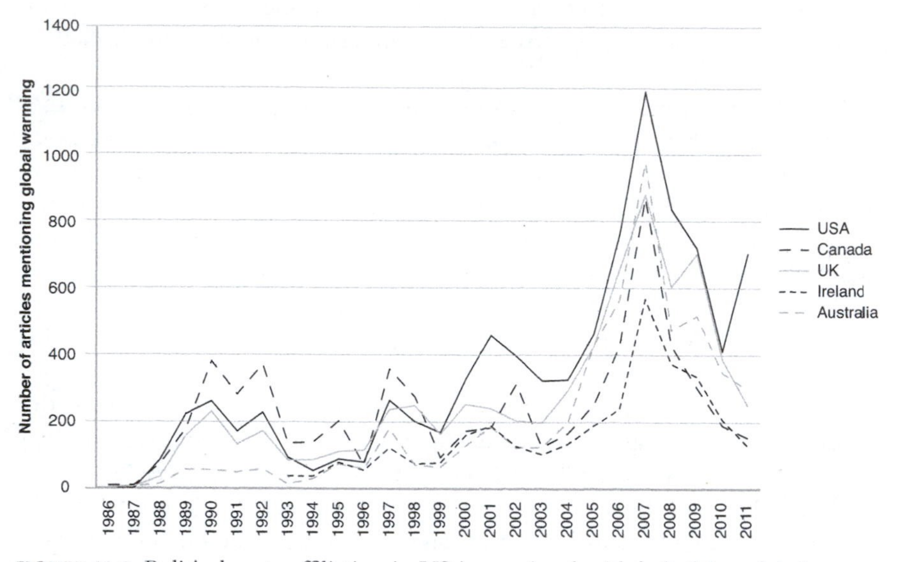

FIGURE 11.2 Political party affiliation in US is associated with belief that global warming is serious (Gallup)

relevance to the technical issue, and it must be politically compatible, or at least not incompatible, with the ideology or political views of the activists.

Conclusion

Complete agreement among technical experts is like the end of the rainbow: You may approach it but never reach it. Science is not like a slot machine, where in one sublime moment four cherries come up and everyone knows that's a winner. Scientists have all the human foibles that create dissenters, including stubbornness, envy, defensiveness, and simple variation in what kind of evidence is required to become convinced of something.

Sociologists entertain essentially two models of how technical knowledge is applied to public policy. The first – call it the "knowledge model" – assumes that scientists *can* obtain approximately true answers to their research questions, with methods that are fairly objective. This knowledge is used to inform public policy. If scientists determined that burning fossil fuel was warming the atmosphere, then policymakers would use this as one factor in deciding what, if anything, to do about it.

The second – or "politics model" – can be applied whatever one's view about the objectivity of science. Here partisans use scientific findings as political capital, to sway policy in the direction they prefer. If you want to curtail fossil fuel consumption, you attribute harmful effects to it; if you want to continue fossil fuel consumption, you claim little or no effect. It makes little difference if your findings are correct, objective, or honest – as long as they are persuasive. Findings that work against your position are buried, or if that is not possible are attacked as invalid or inapplicable. In the politics model, scientific claims are used polemically, just like any other kind of political argumentation (Mazur 1998).

Even experts seem to have difficulty recognizing which of these models is working in a given situation, and laypeople have no chance at all. Highly politicized technical disputes, as they are aired in news media, legislative hearings, or public protests, often seem as if they are seething caldrons of inchoate motives, beliefs, and interests, unfathomable and incapable of resolution. Here is my own stance: When we cannot judge the facts ourselves, and there is a very large majority of scientists in relevant fields agreeing on one factual position in a transparent deliberative process like that of the IPCC, believe it.

12
PEOPLE RESPOND TO NEWS COVERAGE

In the 1950s many scholars believed that newspapers could sway public opinion toward a particular view favored by the publisher, just as advertisements convinced consumers to buy a particular product brand. Perhaps this reflected still fresh impressions of Nazi Germany, where the Ministry of Propaganda would put out a story, often a "big lie," that was obligatorily repeated in the German media and supposedly accepted by the German public. An abundance of modern research has made clear that news organs cannot effectively lead readers or viewers to an intended position (nor are advertisers very effective in persuading shoppers to buy a particular brand, beyond giving it visibility). For one thing, people are selective in their consumption of news, choosing sources and stories that roughly agree with their existing positions, and ignoring or not believing those of opposite persuasion. For another, people are highly swayed by their personal associations – family, friends, and colleagues – far more than by impersonal mass media. Altogether, mass media by themselves do not have much power to push people to positions that they are not open to anyway, and that is probably true of social media too.

Quantity of coverage theory

The most solidly demonstrated effect of news media on opinion is agenda-setting, the placing of certain issues or problems foremost in the minds of people, including policy-makers, simply by making them salient in news broadcasts and publications. As a classic formulation puts it, the news media are not successful in telling us what to think, but they do succeed in telling us what to think about (McCombs & Shaw 1972).

Quantity of Coverage Theory (QCT) builds on agenda setting (Mazur & Lee 1993; Palfreman 2006; Andrews & Carena 2010). As applied to energy or environmental hazards, the theory asserts:

1. People do not usually attend to the detailed content of news coverage; instead they absorb simple images of hazards, like polar bears stranded on melting ice floes as a symbol of global warming, an ozone "hole" over Antarctica, or inextricably linking bombs and radioactive contamination to nuclear power (Weart 2012).
2. People are affected more by the quantity of coverage, especially the repetition of simple images, than by detailed content (the "availability heuristic" [Kahneman 2011]).
3. Public worry, partisan activity, and government action rise and fall with the quantity and saliency of news coverage about a hazard.
4. The quantity of coverage given an alleged hazard is determined more by "externals" – such as the prominence of related issues, and relationships among journalists and their sources – than by authoritative evaluations of the validity or severity of the hazard.
5. Most environmental risk stories of national or international scope are first brought to widespread attention by a small, central group of large news organizations including major newspapers, wire services, and television networks; and by prominent sources including government and environmental spokespeople. Every day these national organizations and influential sources produce a pool of news articles from which thousands of local organs select their news of the day.
6. Therefore the rise and fall of public and governmental concern may be traced back to the rise and fall of coverage by the central media.

Here we examine how QCT applies to accidents, then to the dispute over global warming.

Accidents

Civil engineer Henry Petroski (1992) argues persuasively that accidents are essential for the progress of technology. If there were no failures, engineers could not locate the cusp between bridges that are very strong but too expensive to build, and those that are cheap enough but too weak to carry the load. There would be no pushing the envelope, no risk taking to advance the art. Accidents, properly analyzed, make us safer in the future.

Sociologist Charles Perrow (1999) persuasively claims that the engineering approach to systems as complex and tightly coupled as a nuclear power plant is bound to fail because unanticipated anomalies inevitably occur. As a result, serious accidents are not only a certainty but a "normal" part of operation. Perrow analyzes a range of accidents, but his paradigm case is the 1979 accident at the Three Mile Island nuclear plant in Pennsylvania, where a wholly unanticipated and misunderstood cascade of events – including operator errors due to inadequate training as well as mechanical failures – caused a partial meltdown of the reactor core. Other famous accidents that more or less fit Perrow's pattern are the Chernobyl nuclear disaster in 1986; the loss of space shuttle *Columbia* in 2003; and the 2010 explosion at the *Deepwater Horizon* drilling rig in the Gulf of Mexico (Graham & Reilly 2011).

The postmortem of an accident is intrinsically difficult, physically because much evidence is destroyed, and socially because every party feasibly at fault blames someone or something else. If a triggering event is identified, the forensic analyst can always find additional contributing factors. The space shuttle *Challenger* exploded because a simple O-ring lost its flexibility in cold weather and therefore did not properly function as a seal; that was the proximate cause that led to a fuel leak and explosion. Sociologist Diane Vaughn's (1997) analysis of the *Challenger* accident emphasizes in addition the excessive "risk culture" at NASA that pushed mission controllers to launch despite cold weather and warnings from other engineers that seals might not hold. The *Deepwater Horizon* catastrophe was also blamed, in part, on an inadequate culture of safety by the oil corporation BP and its contractors involved with the rig (Coll 2012).

Some major energy-related accidents are not complex and can reasonably be described as stupid human error, oversight, or irresponsibility. It was a clear night in March 1989 when the *Exxon Valdez* began its voyage on the calm water of Prince William Sound, Alaska. Encountering icebergs in the shipping lanes, Captain Joseph Hazelwood ordered the helmsman to steer out of the lanes to avoid them, not an extraordinary maneuver. Then, ignoring regulations, Hazelwood, apparently under the influence of alcohol, left the bridge and went to sleep in his cabin, instructing the third mate to return to the shipping lanes. The third mate, perhaps overtired, did not steer the ship back into the lanes but instead into Bligh Reef, ripping the hull.

No prominent extenuating circumstances for the improper steering have come to light. The reef was well marked and well known; the ship's navigation equipment was working; it was a clear, calm night. Beside regular radar, there was onboard a specialized collision-avoidance radar that might have warned the mate if it had not been broken, but it was not essential to avoid the reef. Once the accident occurred, leaking oil was not expeditiously contained, dispersed, or cleaned up, as called for in the "fantasy" emergency response plan (Chapter 10), making this the worst spill in American waters until the *Deepwater Horizon* surpassed the record.

The designers of the Fukushima Daiichi nuclear reactor complex on Japan's western coast foresaw the possibility of a tsunami washing over the reactors and built a seawall 5.7 meters high as protection. Like the structural designers of the World Trade Center who recognized the possibility of an airplane impact, they underestimated or ignored the magnitude of what would actually occur. On March 11, 2011, an offshore earthquake, one of the largest ever recorded, broke the complex's connections to the power grid and spawned a monstrous tsunami. The wave, with a height estimated at about 14 meters, easily swept over the seawall, disabling on-site emergency generators intended as backup power for the core cooling systems. As a result, the three active reactors overheated and suffered meltdowns. There followed a days-long cascade of hydrogen explosions and other disabilities that led to a release of radioactive gases and months-long work to stabilize the reactors (ANS 2012). At first glance, the nub of the accident seems simple: an extraordinary tsunami overwhelmed the inadequately protective seawall, disabling the reactors and their emergency cooling systems. But the matter is not settled unambiguously. In July 2012, a Japanese parliamentary inquiry concluded that the accident was a preventable disaster rooted in government-industry collusion and the worst conformist conventions of Japanese culture, which together failed to provide adequate protections against known hazards (Kurokawa 2012).

Engineering versus sociological models of accidents

A forensic engineering investigation would seek the proximate causes of an accident (e.g., a cold O-ring, an overtired or inept third mate, an extraordinary tsunami), as well as precursors that enabled or led to the proximate cause (e.g., an overzealous "culture of risk," corporate policy that did not enforce adequate rest periods for tanker crews, an inadequate seawall). It would assess the damage done by the accident, and draw lessons that might help avoid such accidents in the future: don't launch shuttles in cold weather; closely monitor tanker captains; build higher seawalls. Forensic investigators certainly recognize that in some cases journalists flock to the scene, giving a mishap and its consequences enormous news coverage, but probably they would not regard the journalistic attention per se as part of the accident (Figure 12.1a).

The main difference in the sociological model presented here is an emphasis on the journalists, who to the engineer are epiphenomenal but to the sociologist are intrinsic to the accident *as a societal phenomenon*. Why do some accidents receive heavy news coverage while others are not mentioned in the news? Severity of damage is one factor affecting the amount of news, but as it turns out, not a very important one. Just as there are precursors to the physical failure, there are precursors that affect the amount of media coverage (Figure 12.1b).

In the engineering model, news coverage is epiphenomenal. While an accident is in progress, recovery managers may regard journalists as a distraction if not a nuisance, leaving public relations specialists to "handle" the reporters. In the sociological model, news coverage – produced by the interaction between journalists and their sources of information – is part and parcel of the accident itself, certainly to the extent that the accident raises public concerns about, fears of, or opposition to a technology, and to the extent that policymakers and politicians attend to it, perhaps altering regulations or allocating (or cutting) funding. In Figure 12.1b this relationship is shown with a two-headed arrow because the societal effect reciprocally affects the amount of coverage; if the accident reaps a lot of attention, media outlets will press to find more words to print, people to interview, and pictures to show.

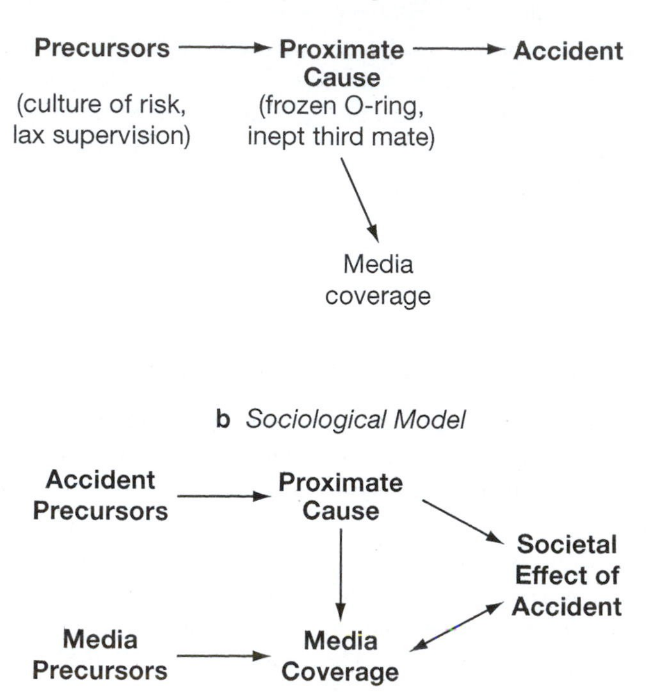

FIGURE 12.1 A AND B Engineering and sociological models of accidents

Nuclear power accidents

Before the Three Mile Island accident in 1979, it was the consistent position of the nuclear power industry and the US Nuclear Regulatory Commission (NRC) that a seriously dangerous failure of a nuclear reactor was so unlikely as to be implausible (e.g., Starr 1969; Rasmussen et al. 1975). Yet there were reactor accidents before TMI and continued to be in the years afterward.

Three nuclear experts whom I interviewed in the mid-1980s, one each from the NRC, the nuclear industry, and the oppositional Union of Concerned Scientists, agreed on the five most serious accidents to date, in the technical sense of potentially leading to harmful events (Mazur 1984; 1990). TMI was the worst. The second most serious, in their judgment, occurred three years earlier, in 1976, at Browns Ferry, Alabama, when workmen started a fire while inspecting for air leaks with a candle flame, destroying wiring in both the redundant control systems; the operators thus lost control of the reactor, though in the end there were no lasting consequences. Another serious pre-TMI accident occurred at Rancho Seco, California in 1978, when an operator who was changing a light bulb dropped it into the control panel, shorting out the power to portions of the plant instruments, which resulted in a loss of monitoring capacity. Both the Brown Ferry and Rancho Seco accidents went unreported by news media.

TMI itself was a dramatic "media event," consuming nearly 40 percent of American television network news during the weeklong crisis, with coverage repeated around the world. (I first heard of it while I was in the Sinai desert.) News attention was especially extensive within the first two days, *before* it was known to be a serious accident. Hundreds of reporters were already at the site when, on the third day, the NRC announced that a hydrogen bubble had built up inside the reactor and might explode, raising the possibly of a breach in the containment, frightening the journalists and the neighboring population (Sandman & Paden 1979). There were urgent concerns about a meltdown and the need to evacuate the population. No one doubted by day three that this was a very severe accident. That night, Walter Cronkite, then dean of TV newscasters, told his audience, "The world has never known a day quite like today." But the story had gone ballistic from day one. Thus, severity alone does not account for the amount of media coverage.

The pattern of attention to the accident is better explained by the increasing interest of editors and reporters in nuclear power during the 1970s, a decade of "energy crisis" and growing opposition to nuclear power. In the mid-1970s, journalists were still insufficiently sensitive or informed to give their attention to the Browns Ferry fire, although the owner of the reactor, the Tennessee Valley Authority, held a formal press conference immediately after the accident.

Perhaps if TMI had occurred in 1976, it too might have been ignored initially – though certainly not after the third day. Coming in 1979, just as President Jimmy Carter was planning a major energy program, and as the anti-nuclear film, *The China Syndrome* (starring Jane Fonda, Jack Lemon, and Michael Douglas), was showing across the country, TMI immediately brought an intense response from news media, some of them juxtaposing the real and movie accidents. Also, unlike Browns Ferry, TMI was in a populous area and easily accessible to journalists in nearby New York and Washington. These were precursors of high media coverage.

As events developed after the second day – the struggle with the bubble, the heroic actions of the nuclear engineers in averting disaster, the arrival of President Carter (a former nuclear engineer) on site, the exodus of the frightened populace – TMI became a first-rank disaster story by any criterion, and with so many journalists already on the scene, attention in the press, radio, and television increased tremendously.

A commission created by President Carter to investigate the accident, more or less following the engineering model, additionally appointed a task force especially to examine the activities of the journalists. Inquiring if the hundreds of reporters who covered the story – usually with little technical understanding of the events before them – acted in a responsible manner, the task force concluded that they had done a creditable job under extremely difficult circumstances. Furthermore, it said that when the news media produced erroneous reports, these errors were usually traceable to sources in the government and industry rather than to the journalists (President's Commission 1979).

Perhaps the most alarming but erroneous report during the accident was that a hydrogen bubble inside the reactor could mix with oxygen and explode, blow the head off the reactor, and release radioactive material into the atmosphere – and that this might occur within two days. The direct precipitant of the story was an NRC employee who telephoned the Washington bureau of the Associated Press to say that hydrogen, which really had accumulated in the reactor, might explode. Over the next few hours, AP reporters worked to confirm and fill out the story, which they were able to do largely through additional sources at the NRC. In the calmer aftermath of the accident, experts agreed that there was no oxygen in the reactor and never was good reason to think that the bubble might explode.

Once the awareness of the press and the public had been heightened by TMI, for a few years afterward, even minor mishaps at nuclear power plants were reported. The terrible accident at Chernobyl in 1986 was of course heavily reported. But by the late 1980s, the theme entered a period of quietude. In 2002 news media barely commented on a near-accident at the 25-year-old Davis-Besse nuclear power plant in Ohio, the most serious at an American reactor since TMI. On March 5, 2002, maintenance workers fixing a leaking tube stumbled on corrosion that had nearly eaten through the top of the six-inch-thick steel reactor vessel, leaving only a half-inch layer of stainless steel to hold in very highly pressurized cooling water. This should have been discovered in routine inspections but was not. The stainless steel was bent and would have broken if the corrosion had continued, spilling thousands of gallons of slightly radioactive and extremely hot water in the containment building. A loss of coolant raises the prospect of core damage and the possibility of a meltdown. Surprised NRC officials had never seen so much corrosion in a reactor vessel.

Although the corrosion did not lead to an accident, the NRC regarded it a serious safety incident, kept Davis-Besse shut down for two years, and imposed its largest fine ever – $5 million – on the plant's operating company for actions that led to the corrosion or failed to discover it earlier. The company paid an additional $28 million in fines to the US Department of Energy (Wald 2002; NRC 2009). In the month after the accident, *The New York Times* ran two articles. Other media outside Ohio barely noticed, and the public remained uninformed (and unalarmed).

Oil spills

Everyone knows about BP's *Deepwater Horizon,* the drilling rig that exploded in April 2011, killing 11 workers and until capped three months later, leaking an estimated 4.9 million barrels of oil into the Gulf of Mexico. It is the largest accidental marine oil spill in the history of the petroleum industry. But who knows about the *second* largest spill, lasting from June 1979 to March 1988, when a blowout at Ixtoc I, an exploratory undersea well of Mexico's government-owned oil company Pemex, spilled three million barrels into the Gulf of Mexico? Prevailing currents carried oil to the Texas coastline. *The New York Times* ran only ten articles on Ixtoc I during the year after the blowout. Pemex spent $100 million cleaning up and avoided most claims for compensation by asserting sovereign immunity as a state-owned company (http://en.wikipedia.org/wiki/Ixtoc_I_oil_spill).

The third through seventh largest spills, occurring between 1983 and 1996, were little or not-at-all noted by Western media though each lost between 1.8 and 2.2 million barrels. They were Third World accidents (three in Africa), usually ignored by news organs of the industrial nations.

When news broke of the *Deepwater Horizon* explosion, ExxonMobil executives publically criticized BP's sloppy management style. Ten days later, a ruptured ExxonMobil pipeline dumped 2,400 barrels in coastal areas of eastern Nigeria, soiling shorelines with seaside villages. In his book about ExxonMobil, Steve Coll noted, "The affected area is far from American television news bureaus, and its kidnapping gangs made it a risky place [for journalists] to travel in any event. The spill barely registered" (2012: 682).

The eighth-largest spill was well reported by the Western press and may be remembered by older readers. In 1978 the American-owned *Amoco Cadiz* ran aground on the coast of Brittany, France. Losing 1.6 million barrels, it was the largest spill at that time. The ship later split into three parts and sank. The site was accessible and familiar to reporters, the accident extraordinary in scope, dramatic and photogenic – all elements of a good news story.

The first heavily reported oil spill disaster was about a decade earlier, the shipwreck of the supertanker *Torrey Canyon* in 1967, polluting the scenic coast of Cornwall, UK with 700,000 barrels of crude. Looking only at large industrial oil spills (10,000 barrels or more) in the neighborhoods of the United Kingdom and the United States, there have since been another seven in (or offshore) Britain, Ireland, and France; and 17 in the United States plus two in Canada (http://en.wikipedia.org/wiki/List_of_oil_spills). I compared British and American press coverage of each spill using *The New York Times* and *The Guardian* (London) as representative newspapers. Articles on each spill are countable in the Lexis-Nexis database or from each newspaper's digital archive. (In what follows, I ignore five American spills caused by Hurricane Katrina in 2005 because immense media coverage of that disaster overwhelms attention to individual events.)

Figure 12.2 is limited to oil spills in the US and Canada. For each spill, the numbers of articles in *The New York Times* and *The Guardian* (London) are shown, one above the other, as a function of the amount spilled (in millions of barrels). The figure shows that if an American or Canadian spill gets any mention in the news, the *Times* carries more articles about it than the *Guardian*. Figure 12.3 is the same kind of display for spills in the UK, France, and Ireland. Now the pattern is reversed: The *Guardian* always has more articles than the *Times* (Lee 1993). This is not surprising; each nation's press gives greater attention to accidents near home than to those far ways.

In each figure I have fitted regression lines to each newspaper's data points, solid lines for *The New York Times* and dotted lines for *The Guardian*. Their upward slopes indicate that amount of spillage has some bearing on the amount of news coverage, but all data points do not sit closely to the lines, so more is at play.

The *Amoco Cadiz* was more highly reported in both British and American news than would have been expected from the regression lines. The American press gave far more coverage to the *Exxon Valdez* than "expected" from the size of the spill. At 440,000 barrels, it was much smaller than Ixtoc I, but in 1989 Exxon's spill was the biggest ever in American territorial waters. Also, perhaps, the timing spurred journalistic attention. The late 1980s were years when American concern about the global environment was in a crescendo. Global warming had become a public issue in the summer of 1988, joining the ozone hole, rain forest destruction, and mass extinction of species on the national agenda of problems. Probably Earth Day 1990 was the largest ever in public attendance. Although the leaking *Valdez* was far from the news centers of New York City and Washington DC, it was an easy helicopter flight for reporters and photographers embarking from Anchorage.

Ease of access and photographic opportunities are important precursors for coverage even within the well-connected industrial world. In 1988 an explosion caused the tanker *Odyssey* to sink, spilling over four times as much oil as would be lost the next year by the *Exxon Valdez*. This was reported in one article in *The New York Times*, none in *The Guardian*. The *Odyssey* went down 1,300 kilometers from the coast of Nova Scotia, not in an ecologically sensitive area nor easy to reach for pictures (International Council for the Exploration of the Seas 1990).

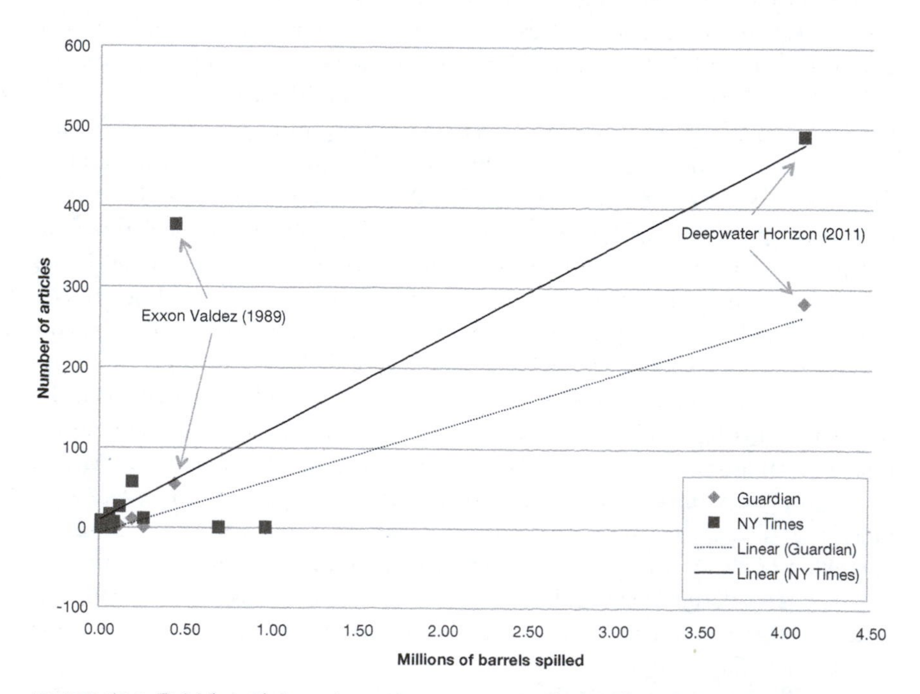

FIGURE 12.2 British and American news coverage of oil spills in North America (Lexis)

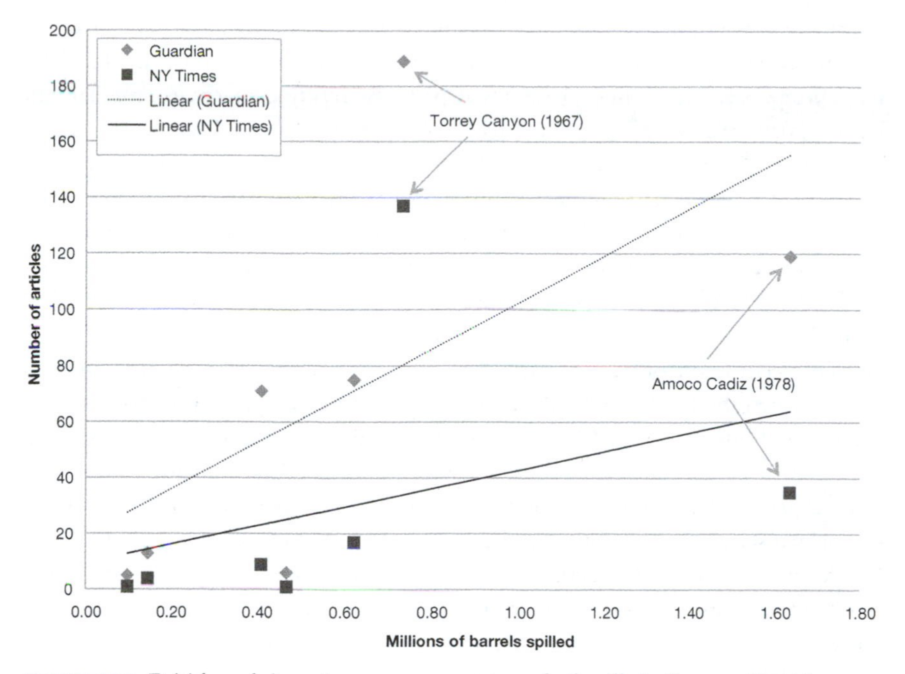

FIGURE 12.3 British and American news coverage of oil spills in Europe (Lexis)

Media coverage of global warming

During the 2008 campaign for the American presidency, Democratic candidate Barak Obama and Republican candidate John McCain both stressed the importance of reducing carbon emissions into the atmosphere in order to limit the impact of global warming. As of this writing in 2012, the Obama administration has produced little progress toward that goal beyond extracting an agreement from US automobile manufacturers to improve fuel economy. In 2012, among the several Republicans vying for their party's presidential nomination, only John Huntsman publicly asserted the reality of global warming, and he dropped out early. Mitt Romney, the eventual nominee, was uncertain; the rest of the hopefuls expressed skepticism or outright denial. Even if they believed it privately, no one stood to gain votes by stressing the need to reduce carbon emissions.

We see the same drop in attention since 2008 in Europe and most of the rest of the world. Why do highs and lows of American attention match highs and lows of public attention elsewhere? Is it because other nations follow America's lead in deciding what problems are important at any given time? I address these questions by examining the quantity of news coverage given to global warming since the 1980s in the United States and four other English-speaking nations: Canada, the United Kingdom, Ireland, and Australia.

At any given time there are numerous potential environmental hazards that risk entrepreneurs try to bring to public awareness through the mass media. The risk entrepreneurs are often professional partisans residing in environmental and health organizations, in governmental and intergovernmental agencies, and in universities (Friedman 1986; Shabecoff 2003; Mazur 2004). Sometimes they are amateurs motivated by community problems, politics, or ideology. Those most seasoned have symbiotic relationships with journalists specializing in environment, health, or science, giving newsworthy material to the reporters, while reporters (and editors or producers) give visibility to the entrepreneurs' issues.

It is difficult placing an issue on the national media agenda, and few threats actually become major news stories. Among those that do, there is little correlation between the level of expertly assessed risk and the amount of media coverage (Singer & Endreny 1994; Mazur 2004). Commentators often cite journalists as the originators or authoritative agents who "construct" a dramatic news story (e.g., Zelizer 1992). More accurately, the root of a growing story is the interplay of journalists and their sources, the risk entrepreneurs, who strive to place hazards in the news and to define their meaning. Van Ginneken (1998) names the White House, an inveterate spinner of news, as the number one newsmaker inthe world.

The global media market is increasingly interdependent and has a shared sense of "newsworthiness," shaping common narratives around the world (Price 2002; Shoemaker & Cohen 2005). American news organs, and secondarily British ones, are the major sources of foreign news for the English-speaking audience, other English-language organs, and probably, with France, for non-English language news outlets as well. The Associated Press (AP), a cooperative news service owned mainly by US newspapers, with 242 bureaus and over 6,000 subscribers in 2005, processing 20 million words daily, is the dominant institution in the world news system. For comparison, the British news service Reuters sends 1.5 million words daily. *The New York Times* sends stories to 130 newspapers abroad, while 300 foreign newspapers subscribe to the Los Angeles Times/Washington Post News Service (van Ginneken 1998; Hachten & Scotton 2007).

The New York Times is often first to nationally publicize particular risks and seems influential in setting the agenda for other periodicals (Krimsky & Plough 1988; Lanouette 1990). Awareness of environmental threats crosses national boundaries. Issues can flow into the US as well as outward. Examples of in-migration are the alarm raised in Britain that childhood vaccines might cause autism, and the famous thalidomide warning, which originated in Germany. Far more often, warnings flow out of the US (Mazur 2009). Foreign editors are more attentive to American news and events than vice versa. The US press gives relatively little coverage even to warnings that are salient in Europe like those over cell phones ("electrosmog") and genetically modified food (Gaskell et al. 1999; Leiss 2001).

Prior work that applied QCT to international news evaluated two competing hypotheses:

Hypothesis 1: Simultaneous peaks in coverage occur in the media of many nations because the central US news organs and their sources influence the news agenda of other countries. Nations most likely to adopt US risk issues are those in closest proximity or with dense trade and communication linkages. Canada is the best candidate, followed by the United Kingdom.

Hypothesis 2: Different national media independently report the same highly newsworthy real-world events, without important influence from US sources.

Each hypothesis may be true, depending on circumstances. The nuclear power plant accidents at Chernobyl and Fukushima Daiichi were sufficiently horrible events to become lead stories in every nation's news media, without any prod from American sources. Hazards like global warming do not cause such terrible short-term events, and these kinds of diffuse risks are especially prone to influence from American news media. Global warming, though surely not purely the domain or invention of Americans, is a case where activities of American journalists and their sources have an outsized effect on rising and falling levels of interest.

To trace year-to-year fluctuations in news coverage in different nations, I selected an important daily newspaper archived in the computer database Lexis-Nexis. *The New York Times* is an obvious choice for the US. *The Toronto Star* is Canada's largest metropolitan daily newspaper. Britain is represented here by *The Guardian* of London. Since 1859 *The Irish Times* has represented the views of the Anglo-Irish community and is now acknowledged as Ireland's leading journal of opinion and information (indexed in Lexis-Nexis only since 1992). *The Sydney Morning Herald* is one of Australia's highest quality and most widely read newspapers.

Searching on "global warming" (in full text) provided counts of all articles by date in the selected periodicals. These were aggregated by year. Trends were checked for consistency with other newspapers in each nation. Alternate search terms ("climate change," "greenhouse effect") produced slightly different counts but similar patterns, an advantage of comparing English-language newspapers, which use similar terminology.

Quantity of coverage in five nations

Theorists since the nineteenth century have understood that atmospheric CO_2 traps heat from the sun, but not until the late 1980s did this become a serious concern. The *Global 2000 Report* commissioned by President Jimmy Carter, which appeared in 1980, considered a future of global cooling as likely as global warming.

Figure 12.4 shows yearly counts of news articles mentioning "global warming" for the period 1986 through 2012 (the last year prorated from the first eight months). There are simultaneous peaks of coverage across nations, first from 1988 to 1992, a second in 1997, a third less distinct peak beginning in 2000, and an enormous rise in 2006 and 2007, falling off after 2008. The last two years of the graph show a departure from this pattern, with coverage of global warming rising in North America but not elsewhere. (These data precede Hurricane Sandy, which hit New York and coastal New Jersey in late 2012.)

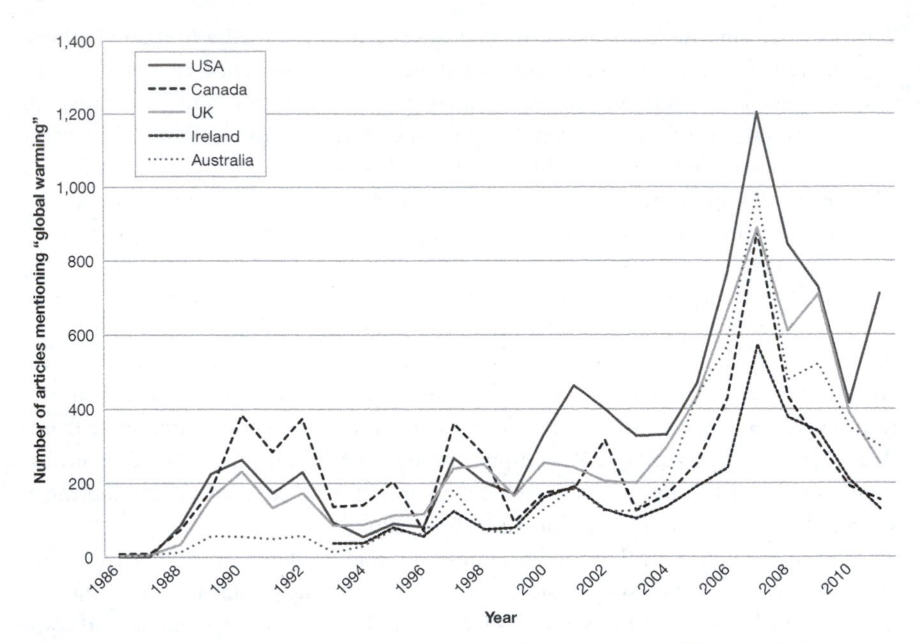

FIGURE 12.4 News coverage of "global warming" in the news of five English-speaking nations (Lexis)

The 1988–92 peak

American concerns about the environment increased sharply after 1986. This upswing was accompanied by a shift in focus from problems that are primarily local or regional in extent (e.g., acid rain, smog, waste disposal) to hazards with worldwide scope, initially the destruction of stratospheric ozone by man-made chlorofluorocarbons, and then joined by destruction of rain forests and mass extinction of species (Mazur & Lee 1993; Lambright 2005).

Interest in greenhouse warming rose along with attention to the ozone hole. Both are problems of the atmosphere, studied by some of the same scientists in the same institutions. In 1988 a US drought, the worst in 50 years, caused major crop losses in the Midwest. New York City, home of most of the national media organizations, was sweltering. The drought became a *Time* magazine cover story on 4 July.

Senator Timothy Wirth of Colorado scheduled hearings on the greenhouse effect for 23 June, the anniversary of the hottest day ever recorded in the capital. The weather cooperated with a temperature that day hitting 38°C. NASA climate scientist James Hansen was scheduled to testify, as he had on three other occasions with little notoriety. But this time the hearing room was full of reporters expecting an important story related to the drought. That evening NBC television news showed Hansen's statement that the greenhouse effect is probably causing global warming now, and the next evening's broadcast connected the greenhouse effect with the drought. (Later, Hansen agreed that long-term climate change cannot be blamed for one season's anomalous weather [Schneider 1989].)

Major newspapers carried the story – on page one in *The New York Times*. *Newsweek's* cover of 11 July, "The Greenhouse Effect," was pegged on the drought and Hansen's assertions. News reports during the summer of 1988 about hypodermic needle-polluted beaches on the East Coast and a massive forest fire in Yellowstone National Park amplified the image of intense heat.

During the summer of 1988, the skyrocketing coverage of global warming carried with it stories about man-made fires during the Amazon dry season, used to clear sections of rainforest for planting. Reporters flew to the Amazon to film the conflagration. (Amazon fires were more extensive the prior year but went unreported in the US.) These were juxtaposed against the huge fire in Yellowstone. Appreciating the press attention in 1988, biodiversity activist E. O. Wilson commented to me, "It's a pity Yellowstone could only burn once."

The clustering together of global environmental problems was by then common and received coverage in news media round the world. George H. W. Bush, during his 1988 presidential campaign, announced that he would be an environmental president. *National Geographic* magazine, for its final cover of 1988, featured a hologram of a crystalline "fragile earth" being pierced by a bullet. (The back of the magazine carried a hologram of McDonald's, sponsor of this extraordinary cover.) *Time* magazine, instead of naming its usual Person of the Year for 1988, featured "Endangered Earth" as its Planet of the Year (2 January, 1989). The Exxon Valdez oil spill of March 1989 drove environmental attention still higher. This crescendo of media coverage and public concern reached a climax on Earth Day 1990, the most widely celebrated ever.

By 1992, US press coverage and public concern were waning, even as leadership of the United States passed to President Bill Clinton and his environmentalist vice president, Al Gore. The sudden outbreak of the Gulf War in 1991, despite publicizing oil well fires started by Iraqi forces, seemed to break the flow of stories on the global environment, excepting a brief revival during the Rio "earth summit" of 1992. The collapse of the Soviet Union was a paramount story occupying news organs. Global temperatures fell in 1991 and 1992 because of sulfate aerosols produced by the Pinatubo volcanic eruption. These factors contributed to the expiration of the endangered earth as a news story.

The 1997 and 2000 peaks

American sources and media played no special role in these intermediate peaks. Unusually high coverage in 1997 is concentrated in the final months of the year and fully explained by the world's nations meeting, first in Bonn, then Kyoto, to produce a protocol calling on industrial countries to reduce their carbon emissions below 1990 levels by the year 2012.

Coverage was again high in 2000, especially during an international conference at The Hague in November to solidify implementation of the Kyoto protocol.

This meeting was newsworthy because of its contentiousness, with the United States, still under the Clinton administration, as well as Canada and Australia resisting a strong regimen supported by the European Union. The conference ended in failure.

The 2006–07 peak

Al Gore is often cited for his contribution to the 2006–07 revival of public concern. The American debut of *An Inconvenient Truth* at the Sundance Film Festival in January 2006, then its European premiere at the Cannes Film Festival in May, received international news coverage, as did the film's two Academy Awards (February 25, 2007), and the announcement in October 2007 that Gore was a recipient, jointly with the UN Intergovernmental Panel on Climate Change (IPCC), of that year's Nobel Peace Prize. But it was the film's opening in theaters nationwide (May 2006 in the US and Canada; September in Britain, Ireland, and Australia), accompanied by release of a book of the same title (Gore 2006), that pushed news coverage to sustained heights, increasing monthly counts of news stories mentioning global warming by one-third to one-half.

American and foreign reviewers contrasted the warm and wise Gore-as-lecturer with the wooden candidate of the 2000 presidential campaign. Rising unpopularity of the Bush administration, domestically and internationally, fueled attention to the man who now introduced himself saying "I used to be the next president of the United States." In 1997, when the peripatetic Gore visited every nation in this study to promote *An Inconvenient Truth*, local newspapers bumped up coverage of global warming.

Without denying Gore's success as a risk entrepreneur, his effect may be overstated. There were other contributors to public concern, not least the objectively rising temperature of earth's atmosphere. Eleven of the years from 1995 through 2006 were among the 12 warmest of the past 150 years. The IPCC's fourth assessment report, released in 2007, affirmed the human contribution to the warming. US gasoline prices were higher in 2006–07 (in constant dollars) than they had been since 1980 in the aftermath of the revolution in Iran. Commentators such as Thomas Friedman of *The New York Times* emphasized the self-destructiveness of sending petrodollars to the Middle East, financing nations whose citizens had attacked the United States. These issues appear abundantly in English-language news coverage of global warming during 2006 and 2007.

However, none of these factors provides a general explanation for peaks of media interest. Gore was not an important element in earlier periods of intense coverage. Media peaks did not occur in record setting years for global temperature, which according to NASA data were 1998, 2005 and 2007. The IPCC's four assessment reports, each increasingly grave about the prospect of warming, were released in 1990, 1995, 2001, and 2007, so none can be the initiator of any media peak. The cost of gasoline was not especially high at the inception of earlier peaks of coverage.

The 2006–07 reporting about greenhouse warming was framed by the industrialized West's enrichment with petrodollars of undemocratic, terrorist-breeding nations of the Persian Gulf. Also present in the coverage were corollaries of warming, such as drought (ongoing in the American South and Southwest, and Australia), melting Arctic ice (with stranded polar bears), and more hurricanes like Katrina. (In fact, the 2006 and 2007 hurricane seasons were unexpectedly mild.) This differed from the frame of 1988–92 when greenhouse effect was one element of a suite of problems endangering the earth, along with ozone depletion, rain forest destruction, and loss of biodiversity. These were little mentioned in 2006–07.

A broader context for global media

The prominence given to American-generated warnings by other nations must be placed in context. The United States is the dominant world power, receiving the lion's share of foreign coverage by other nations over a wide range of cultural, political, and economic topics.

To measure this dominance, in May 2008 I counted in each target newspaper the number of articles (during the month past) mentioning each of the five nations. The *Guardian*, for example, had 1,143 articles mentioning "United States," 264 mentioning "Ireland" (including Northern Ireland), 212 mentioning "Australia," 92 mentioning "Canada," and 4,736 mentioning its own nation, the United Kingdom. The *Guardian's* degree of interest in the United States can be quantified (and standardized) as its number of articles mentioning the US, divided by its number of articles mentioning the UK. This quotient = 1,143/4,736 = 0.24. The *Guardian's* quotients for other nations are considerably lower, indicating less attention paid to those nations than to the United States. (The four *Guardian* quotients are 0.24 for the US, 0.06 for Ireland, 0.04 for Australia, and 0.02 for Canada.) Each newspaper paid more attention to the US than to other foreign nations. One exception is the *Irish Times*, which gave more of its foreign coverage to the UK (quotient = 0.35) than to the US (quotient = 0.24).

To avoid a blizzard of numbers, the pattern of quotients is shown in Figure 12.5. (Quotients lower than 0.04 are omitted.) Arrows point from the reporting nation to the nation being reported (more exactly, to the nation mentioned). This diagram shows that other nations pay far more attention to the United States than US media pay to them.

We must add to this asymmetrical picture the large number of US-produced articles, television programs, and movies, carrying entertainment and news, which are consumed in other nations far more than their products are consumed in the United States (Hachten & Scotton 2007). The high priority accorded American-generated warnings about health and the environment is simply one aspect of this broader pattern of American-dominated dissemination.

News coverage and public opinion

QCT asserts that public worry rises and falls with the quantity of news coverage about a hazard. We may explore this connection within the United States by

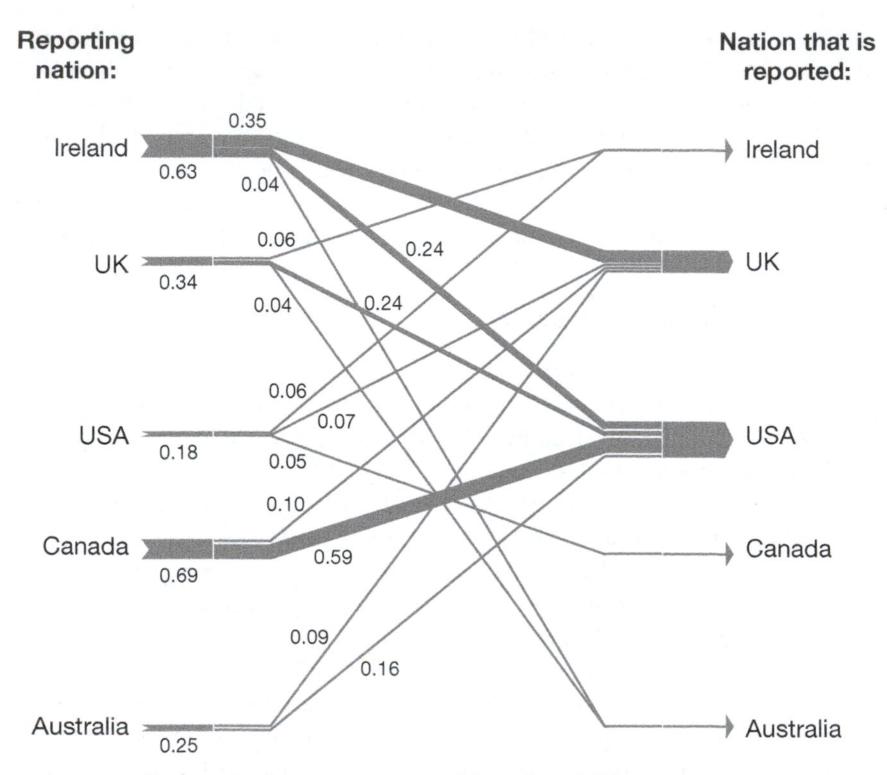

FIGURE 12.5 Each nation's news coverage of the others (Wilcoxen)

comparing the amount of US news coverage of global warming with year-to-year changes in public worry about climate change. The Gallup Poll has the longest time series, recording from 1989 to 1991, and again from 1998 to 2012, the percentage of Americans who say they worry "a great deal" about global warming. These percentages are shown as dark bars in Figure 12.6 (http://www.gallup.com/poll/126716/environmental-issues-year-low-concern.aspx). The highest level of worry was in 2007 when 41 percent of respondents said they worried a great deal. (In years when Gallup did not ask this question, pollsters likely thought that global warming was of minor concern.)

The light bar in Figure 12.6 shows an index of news coverage. (This is the yearly number of articles in *The New York Times* mentioning global warming, divided by 22, the reason being to show the light and dark bars with comparable vertical scaling.) Visual inspection shows that news coverage (light) and public worry (dark) rise and fall together.

Conclusions

Rises and falls of news coverage about global warming were remarkably synchronous in all five English-speaking nations. Greenhouse warming received

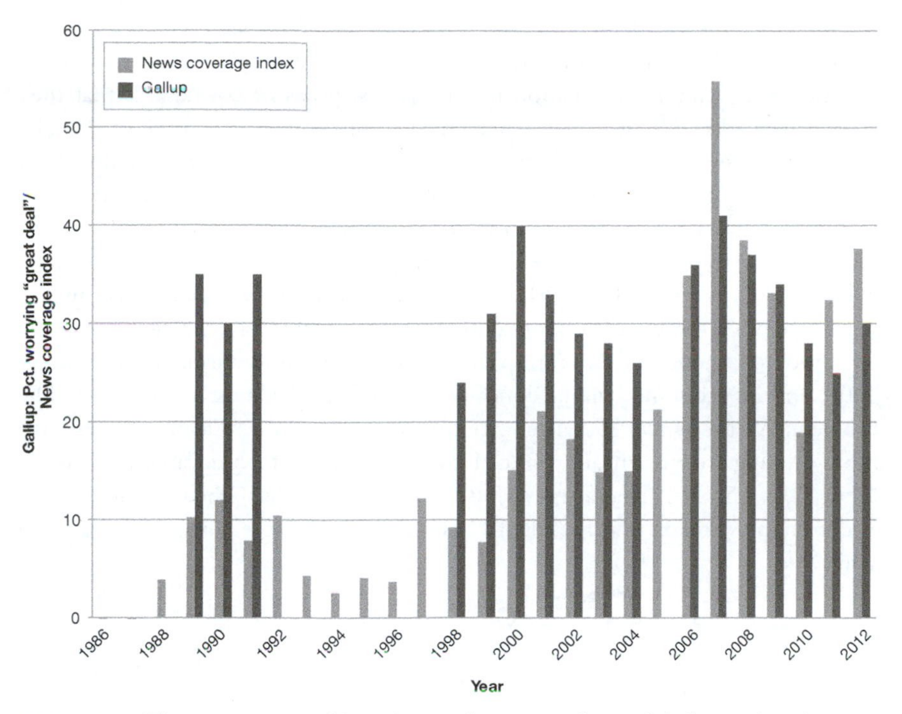

FIGURE 12.6 The percentage of Americans who worry about global warming rises and falls with the amount of news coverage of global warming (Gallup, Lexis)

worldwide attention during the period 1988–92 as part of a suite of global threats including stratospheric ozone depletion, rainforest destruction, and mass extinction of species. After 1992, coverage of these problems fell away, despite the continuing destruction of rainforests and loss of species, and the still-rising temperature of earth's atmosphere. There was, beginning in 2006, a huge revival of news about global warming. This up-and-down pattern of news coverage is unrelated to changes in the evaluation of the threat by experts.

The 2006–07 peak is closely associated with Al Gore's documentary, *An Inconvenient Truth*. Gore's warning about this problem, in sharp contrast to then-President George W. Bush's refusal to sign the Kyoto Protocol, may partly explain the polarization in views about climate change between Republicans/conservatives and Democrats/liberals (Chapter 11). One easily forgets that environmental protection began as a bipartisan issue.

It is impossible to definitively establish with correlational data the cause of simultaneously fluctuating media coverage, or to prove an American origin for sharp peaks. The best one can do is evaluate the plausibility of alternative explanations, most particularly that journalists in each nation independently report especially newsworthy events (Hypothesis 2). This sometimes occurred, as in the small peaks of 1997 and 2000 about international conferences in Kyoto and The Hague, but peaks are not generally initiated by a clearly recognizable real-world manifestation of the threat. Worldwide fluctuations in coverage of global warming are especially hard to reconcile with the continuous heating of the atmosphere.

The most plausible explanation for the largest peaks of coverage is that they were generated by US media or American risk entrepreneurs (Hypothesis 1). The greenhouse story's sudden rise to prominence in August 1988 was due to a constructed media event, the testimony of NASA scientist James Hansen before Congress, when he (unjustifiably) linked the summer drought to climate change. Journalists were alerted and present, primed by the summer heat wave. The 2006–07 peak is largely the work of Al Gore. But the news media have limited space, and other stories vie for journalistic resources. Obviously, the global financial crisis took over much of the front page, especially during the panic of September 2008, and its economic fallout continues to rank as the most important public problem. All the while, global temperature moves upward, but most of us have more pressing worries on our minds. Hurrican Sandy, which devastatingly struck New York City and coastal New Jersey in late 2012, brought the attention of American news media and President Obama back to climate change, at least for a while.

PART V

Progress and regress

13

ATTEMPTED SOLUTIONS

"Drill, baby, drill!" has become a rallying cry for politically conservative ideologues who believe the solution to our energy problems is to extract and burn ever more fossil fuels. On the opposite extreme are calls to eliminate fossil and nuclear energy entirely and immediately, wholly replacing them with renewables and conservation. Both nostrums are simplistic, unrealistic, and ill informed. But what are more cogent remedies?

Real solutions face technical and sociological barriers, they must be affordable and politically palatable, and they must survive resistance from the social institutions devoted to increasing energy consumption. In this penultimate chapter, we look at a handful of approaches that have been tried sufficiently that we can draw lessons from them.

Cap and trade and the Kyoto Protocol

Modern antipollution regulation is usually dated to 1970 when the US Congress passed a strong Clean Air Act intended to reduce six "criteria" air pollutants commonly emitted from industrial and mobile sources and injurious to health. (CO_2 was not then regarded as a health-damaging pollutant.) The US Environmental Protection Agency was created to oversee and enforce limits on pollution. Its approach has been called "command and control" because the EPA set a maximum permissible limit on each pollutant. Any emitter, e.g., a power plant, exceeding that limit was penalized. The new law (with subsequent amendment) was most successful with airborne lead and carbon monoxide, less with ground-level ozone, but overall it is credited with reducing air pollution in the US even in the face of rising population, a growing economy, and more vehicle miles driven (www.epa.gov/airtrends/). Other industrial nations adopted similar laws (as did some Third World nations including China, though without serious enforcement).

Business interests complained with some justification that command-and-control was inflexible and could be an excessively expensive way to reduce pollution. For example, a power plant with emissions barely over the permissible limit might require very expensive modifications to reach compliance. Political conservatives objected to the heavy hand of government regulators. Two alternative approaches were discussed to reduce pollution. One was to charge a pollution tax, i.e., the emitter would pay to the government a tax determined by the amount of noxious material it put into the environment. This would allow emitters to balance reduced emissions with increased taxes in any way that was advantageous to them. Since tax collection is a routine function of government, it would not require an extraordinary increase in the bureaucracy. The difficulty with this approach is that politicians do not want the unpopularity that comes with imposing new taxes.

Conservative economists pointed out a second alternative called "cap and trade," whereby Congress could avoid the onus of passing a new "tax" yet produce essentially the same effect. Imagine a city entirely covered by a huge glass dome. Within the city are several major polluters that together emit 10,000 tonnes of SO_2 per year. Say your goal for the next year is to reduce ("cap") that to 9,000 tonnes (with more reductions in later years). You print 9,000 permits, each allowing the holder to emit one tonne of SO_2. These are given or sold to the city's polluters, perhaps distributed according to the size of the enterprise or its current level of emissions. Polluters can now choose any course they like as long as they do not emit more tonnes of SO_2 than they have permits for. One polluter might invest in a new facility that emits less noxious emissions, another might install pollution control equipment on an existing plant, and another might close a very dirty plant and shift operations to a less polluting facility. Each operator can choose whatever course seems best for them. Some will cost-effectively reduce their tonnes of SO_2 emissions to the number of permits in hand. Some will do even better, finding that they hold more permits than they need for next year's operations. Here is where "trade" comes in. Holders of unneeded permits can sell them to other polluters who have insufficient permits to cover their expected emissions. This is accomplished by setting up a market where SO_2 permits are traded, like a stock market, the price of a permit determined by supply and demand. Large polluters could continue their dirty operations unabated and without penalty as long as they bought enough additional permits.

Critics immediately objected that cap and trade gave polluters permission to pollute. This is true, but looking at the overall effect on the dome-enclosed city, SO_2 emissions are reduced from 10,000 tonnes to 9,000 tonnes, and this is accomplished without levying a new tax and while businesses control their own affairs. The idea gained sufficient force that it was incorporated on a trial basis into the 1990 amendments to the Clean Air Act, specifically to reduce SO_2. This trial was remarkably successful, lowering SO_2 emissions farther and faster than predicted, and without much complaint from corporations or politicians. It worked so well because SO_2 is produced mostly by a few large stationary sources like coal-fired power plants, so it is relatively easy to monitor emitters and to set up permit trading.

Also, there were technologies already operating that could be conveniently used to lower SO_2, including dedicated trains that inexpensively deliver low-sulfur coal from Wyoming to eastern power plants, and scrubbers that remove SO_2 from stack gases. Enthusiasts who then sought to use cap and trade to control CO_2 emissions faced a harder task because important sources of emission, including vehicles, are far more numerous and dispersed, and it is not so easy and inexpensive to capture and store CO_2 on a large scale. Perhaps most important is that greenhouse gases must be controlled not just on a regional level, like SO_2, but on a global scale, requiring cooperation among nations that are significant emitters.

The UN's Kyoto Protocol was adopted in Kyoto, Japan in 1997, its intent to forge an international agreement to reduce greenhouse gases (GHG) using cap and trade as one mechanism to achieve that end. Although China is presently the largest emitter of CO_2 (the US is second), Third World nations were exempted from Kyoto obligations on the principle that the industrial world is mostly responsible for today's high level of atmospheric CO_2, and developing nations should be allowed more slack in using fossil fuels to pull themselves out of poverty. For 37 industrial nations that ratified the Protocol, it went into force in 2005, binding them to GHG reductions by 2012 that on average would bring emissions down four to five percent below 1990 levels. Reduction targets varied from one participating nation to another, depending on their circumstances and negotiating abilities (http://unfccc.int/Kyoto_protocol/items/2830.php/).

The United States did not ratify the agreement, an abdication of responsibility often blamed on President George W. Bush but already obvious during the prior Clinton-Gore administration when a bipartisan Congress almost unanimously opposed the treaty because of its unknown effects on the American economy and China's exemption. Canada did ratify the Protocol but failing to approach its reduction target, withdrew in 2011.

While the primary expectation of the Protocol was that each participating nation would actually reduce its own GHG emissions, there were additional mechanisms – critics might call them "escape clauses" – for meeting their targets by 2012. One was cap and trade, called "emissions trading" in the Protocol, whereby one participating nation that had not reduced its GHG emissions can buy emission credits from another participating nation that has made reductions. Another mechanism allowed a participating nation to offset some of its own GHG emissions by potentially reducing GHG emissions in a developing nation, for example, by protecting a tropical rain forest that might otherwise have been cut down. One more mechanism allowed a nation to offset its own emissions by investing in emission reduction in another participating nation, for example, replacing an old coal-fired power plant in a formerly Communist nation with a more efficient gas-fired power plant (Nordhaus & Shellenberger 2007).

Reducing GHG emissions from the base year of 1990 was especially favorable to nations emerging from Communist control because industrial operations in the Soviet Union and its satellites were highly inefficient and could hardly escape improvement under capitalism. Germany, after reunification, was committed to bringing the former East Germany up to the standards of West Germany; satisfying its Protocol target was incidental to the project of national integration. The former satellites or republics of the defunct USSR improved their carbon efficiencies as they integrated with the European Union.

It is premature to give a final accounting of the Kyoto Protocol. By way of a progress report, we can compare recent with past CO_2 emissions in 29 of the countries that ratified the Protocol in 2005 (BP 2012, excludes the smallest nations). Sixteen out of 29 reduced CO_2 output from 1990 to 2011. This includes all nine former members of the Soviet bloc, with greatest reductions in Ukraine and Lithuania (–57 percent and –56 percent, respectively). These reductions are likely due to post-Soviet collapse or conversion to Western technology. Reunified Germany, as expected, also reported considerable improvement (–22 percent). Among nations without a communist background, the largest reductions in CO_2 were reported by the UK (–18 percent), perhaps because of the collapse of Britain's coal mining industry; Denmark (–16 percent), which has made a heavy commitment to wind power; and Sweden (–14 percent), heavily invested in hydroelectricity. France, Switzerland, and Italy reported smaller reductions (<10 percent). On the other side, Spain was emitting 48 percent more CO_2 in 2011 than in 1990, Australia's emissions increased 42 percent, Ireland was 37 percent higher, Canada increased 26 percent, and even energy-conscious Japan was up 12 percent. The European Union overall reduced CO_2 by 10 percent. The US, which did not ratify Kyoto, was up 11 percent.

It is likely that in the final accounting of the Kyoto Protocol, nations that did not reduce their own GHG emissions will invoke allowable offsets to approach if not meet their targets. To date there has been no success in improving or extending the Protocol beyond 2012. In any case, without participation of the largest GHG emitters, the US and China, whatever Kyoto participants accomplish, even meeting their goals, will have little effect on the atmosphere's total load of greenhouse gases by mid-century. Between 1990 and 2011, China increased its CO_2 emissions by 376 percent, India by 309 percent, and there is no sign of diminution.

Do cities save or waste energy?

I have two books on my desk that seem at first glance to give contradictory views of cities in industrial nations as users of energy. David Owen opens *Green Metropolis* (2009) lauding Manhattan as a model of energy efficiency. When he and his wife lived there as newlyweds, they consumed roughly 4,000 kilowatt-hours of electricity per year. Now residing with two children in a small Connecticut town, where most chores require driving somewhere, they put 20,000 miles per year on their three cars and consume almost 30,000 kilowatt-hours of electricity.

The keys to Manhattan's efficiency are its density (26,939 people per square kilometer) and compactness, good mass transit, and ubiquitous taxis. With the proximity of residences to shops and places of work, there is little need for a car except to get out of the city. Cars are actually a liability given the limited and high cost parking, and congested traffic. About three-fourths of Manhattan households do not own an automobile, compared to 8 percent nationally. Many who have cars use them mostly on weekends. Eighty-two percent of employed Manhattan residents travel to work by public transportation, bicycle, or on foot, which is ten times the rate for all Americans. The only US cities that even approach this are Washington DC, Boston, San Francisco, Philadelphia, and Chicago (http://en.wikipedia.org/wiki/Transportation_in_New_York_City).

Owen notes that the average New Yorker (in all of the city's five boroughs) generates less greenhouse gas than residents of any other American city, only 30 percent of the national average, with Manhattanites doing even better. Most residents of the Island live in relatively small apartments in large or high-rise buildings, not in single-family houses, because land in Manhattan is scarce and rents are expensive. Apartments use relatively little heating or cooling because they are isolated from the weather on most sides by other residences. In winter, the heat that escapes one apartment warms those adjacent, or above or below; in summer the air conditioning is effectively shared.

Recall that a thermal electric power plant typically wastes two-thirds of the fuel used to generate electricity. Cogeneration is the process of putting that potentially wasted heat to good use, usually by supplying steam for space heating or industrial processes. New York City's utility company, Consolidated Edison, operates the largest cogeneration service in the United States, using underground pipes to send steam to buildings in the lower two-thirds of Manhattan, another huge energy saving. Apart from Manhattan, cogeneration is surprisingly little used in the United States, less than in Europe, but it could supply heating in many areas of dense settlement.

Owen's optimistic picture of Manhattan is incomplete. He takes no account of the flux of commuters in and out of the city each weekday, usually one person per car, often stuck in traffic with engines idling. This is closer to Austin Troy's take on urban energy metabolism in his book, *The Very Hungry City* (2012). Troy opens with a sketch of the interchange for Interstates 5 and 210 in the city of Sylmar, just outside of Los Angeles. "For those who love the sight of acres of reinforced concrete and steel, this is the place to be. Here, I-5 expands to up to fifteen lanes in width to accommodate the massive amounts of traffic." To me, Los Angles epitomizes urban sprawl, built as if the planet's dominant species were automobiles. When I lived there in the 1960s, one could barely locate a downtown or urban core. Buildings and roads just spread out to the Pacific beaches in the west and to the mountains in the north and east that rim the Los Angeles Basin. Many Angelinos live in sizable single-family houses, the population density far less than Manhattan. Shopping and work are rarely within walking distance, so cars are essential, and the area is famous for its freeways. A new subway system barely compares to New York transit's reach or ridership. Fortunately the warm climate relieves the city of much need for heating.

Fresh water is Los Angeles' really scarce resource. It is expensive to desalinate ocean water so the city depends on usually plentiful snowmelt from the high Sierra

Nevada, flowing by gravity to irrigate California's fertile Central Valley. Carrying the flow farther to the city, especially over the Tehachapi Mountains, requires very large pumping stations that annually consume three-quarters as much electricity as is used by all residences in the city of Los Angeles (Troy 2012: 39).

Actually, authors Owen and Troy are in considerable agreement about what makes some cities energy efficient. In densely populated areas where places for living, working, and shopping are within walking distance, and with good public transit for longer trips, there is little incentive (and often disincentive) for car ownership. Where residential and commercial buildings are compact and clustered, they consume relatively little energy. Manhattanites are not particularly ecologically minded. Their conservation follows from a fortunate mix of urban arrangements with economic incentives and disincentives. In many of America's cities, even the Greater New York metropolis that surrounds Manhattan, large single-family houses are common outside downtown; cars are needed to go to work, for shopping, and to bring children to their various activities; stores with large parking lots sprawl along miles of highway.

Objectively comparing the per capita energy consumption of different cities, or of cities versus rural areas, is plagued by ambiguity in what to include in a city's energy consumption, and how to define a metropolitan area (Sovacool & Brown 2010). Should we count the gasoline used by people who commute to the city for work, then return to their outlying homes at night? Should we count energy consumed in surrounding suburbs or just the city core? How do we tally the energy used to bring freight into or out of the city, to bring in water and other resources, and to remove effluvia produced in the city?

An ambitious report comparing the carbon footprint – a proxy for fossil fuel consumption – of America's 100 largest Metropolitan Statistical Areas (MSAs) is instructive not only for its results but for its methodological difficulties (Brown et al. 2008). Given limits on available data, each MSA's energy consumption was estimated from annual mileage traveled within the MSA for passenger and freight vehicles, and from energy consumed by residential buildings. Energy used by commercial and other nonresidential structures was not included, nor was mileage driven by suburban commuters before entering the MSA. Within these limitations, the authors found that the 100 largest MSAs, which house two-thirds of the nation's population and account for three-quarters of economic activity, emitted just 56 percent of US carbon emissions from highway transportation and residential buildings in 2005. Residents of the 100 largest MSAs had a nearly 15 percent smaller carbon footprint than the US average. On that basis, cities look good for energy conservation.

But there is great variation in carbon footprints among the 100 MSAs. The worst, Lexington, Kentucky, produced 2.5 times as much carbon per capita as the best, Honolulu, Hawaii. No doubt the felicitous climate of Hawaii, requiring little space heating or cooling, and its limits on places where one can drive, contributed to this difference. The New York MSA had the fourth best carbon footprint in the nation, no surprise to Owen or Troy. What might have shocked them was Los Angeles ranking even better, second only to Honolulu. Marilyn Brown of Georgia Tech University, lead author of the report, told me this finding raised questions when first published. One part of the explanation lies in the low heating needs of homes in Los Angles. Another lies in the defined area of the Los Angeles MSA, limited to Los Angeles County and Orange County (barely including Sylmar near its northern boundary). The Greater Los Angeles Area, a term used by the US Census for the Combined Statistical Area (CSA) that additionally includes San Bernardino, Riverside, and Ventura Counties, is more sprawling, has less concentration of core areas, and lower population density. (In 2010, the number of people/km^2 in the two-county Los Angeles MSA was 1,025, while in the five-county Greater Los Angeles CMA it was only 203.) Also, no account was taken of that electricity needed to bring Sierra Nevada water to the city.

A fair summary seems to me to be this: In compact and densely populated cities where destinations are often within walking distance, where there is good public transportation, and where most housing is in large buildings of small apartments, residents use less energy per capita than people living in dispersed communities of large, single-family housing who are dependent on automobiles.

European cities appear to be more energy conserving than most American cities. Being older, their streets are not as suited for automobile traffic; their homes, stores, and shops are in closer proximity, better adapted to pedestrian traffic; they often have superior and more widely used public transit. There are cultural differences too, with Europeans more willing to walk, less insistent on low gasoline taxes, and used to smaller cars and homes. Energy efficient Tokyo has similar characteristics: high population density (8,000 people/km^2), a complex network of very narrow roads that discourage automobiles so there is very low gasoline consumption, a good mass transit rail system, a tradition of walking and bike riding to transit terminals, compact (though single-family and low-rise) housing, and many small shops scattered near residences (Aoki & Aoki 2010; Sovacool & Brown 2010; Troy 2012).

Few cities in America, or even in Europe, match Amsterdam and Copenhagen for biking. Austin Troy credits Copenhagen with the most advanced bicycle infrastructure in the world; perhaps Amsterdam is second. Ninety percent of Copenhageners own a bike (only 53 percent of households own a car); 37 percent commute to work by bike, somewhat less on rainy days. In winter, snowplows clear the bike lanes before clearing the roads. To reinforce bicycling and public transit, Denmark imposes a 180 percent tax on car sales.

Urban energy consumption is conflated with automobile congestion, which would be noxious even if cars ran energy free. The one advantage to congestion is that it encourages people to use mass transit. Any improvement in roads or parking, intended to reduce congestion, encourages more people to drive.

London, Stockholm, and Milan use an approach called "congestion charging" to protect their city centers from crushing traffic. Cameras record license plates of vehicles entering and leaving the protected zone, automatically charging a fee to the driver's account. In 2012 the fee for a nonexempt vehicle entering central

London during business hours was £10 per day. This appears to reduce traffic as intended, but the degree of reduction is controversial. There have been complaints that the fee increases parking congestion just outside the zone, and those businesses within the zone have lost sales, though presumably to the benefit of businesses outside the zone. Congestion charging favors the wealthy, who most easily afford the fee, and might be improved by scaling the fee to the vehicle owner's wealth. A 2007 proposal for congestion charging to relieve Manhattan's traffic was highly controversial and finally killed by the New York State legislature.

Many cities in Europe and America have converted dense shopping areas into attractive pedestrian malls where cars are prohibited. Usually trams, buses, or electric carts are available for people who cannot easily walk. Apparently these are widely appreciated, though no doubt there are some disadvantages. Conceptually, much larger zones could be established where personal vehicles were limited to something approximating a golf cart in size, weight, and speed but more crashworthy. This would maintain much of the convenience and comfort of a car while reducing the weight of the vehicle to only a few times the weight of passengers, offering considerable savings in energy, pollution, and congestion. Since automobiles and SUVs would be excluded from the zone, there would be no danger of one colliding with these light vehicles. Traveling at limited speeds, say 20 kilometers per hour (under 13 miles per hour), and with low mass, collisions of one light vehicle with another, or with a bicycle or pedestrian, would not be extremely damaging yet still allow travel times comparable to those of taxis in rush hour traffic. The zone boundary would be defined to maximize the number of trips that begin and end in it, covering much of a densely populated area. Residents owning or renting light vehicles might use them to travel between zones atop ferries running along railroad tracks, rivers, or interstate highways.

Efficient transportation

A friend who lived in Bethesda, Maryland, a suburb of Washington DC, would on a typically muggy summer morning finish his breakfast, enter the car in his attached garage, drive 45 minutes to the garage of his office building, then take an elevator up to his desk, all without leaving air conditioning. There is no denying the convenience of automobiles that keep us comfortable in any weather, have ample passenger- and load-carrying capacity, securely enclose us in a protective sheath, run on our own schedule, and take us quickly (assuming no congestion) to exactly where we want to go (assuming adequate parking). Today's remarkable machines, affordable by nearly every citizen of an industrial nation, run 150,000 miles (240,000 kilometers) with routine maintenance and go a long way on a tank of gasoline, which in the US costs a half to one-third what is normal in other industrial nations due to differences in government subsidies and gasoline taxes. I like bicycling and do it nearly every day, even in winter, but in my city of Syracuse, New York, my wife and I could not get along without at least one car.

If we have the time, we prefer driving across country to flying on cramped airliners with their security lines, constraints on what we can bring, strict schedules (often delayed), rushing to make (or miss) connections, occasionally lost luggage, and getting to and from the airport. In a car we cruise along interstates, sometimes minor highways for variety, listening to audiobooks, timing ourselves to avoid rush hour traffic but otherwise unconstrained by schedule, picnicking along the way, spending nights at clean, reasonably priced accommodations or visiting friends, enjoying the Western scenery and national parks.

For many it would be a deprivation to surrender their automobiles for mass transit, bicycling, or walking. It is easy enough to envision social arrangements where getting rid of a car would not cause one's lifestyle to deteriorate, where cars are more nuisance than convenience, as in Manhattan, Copenhagen, and Tokyo. The desirability of owning an automobile rather than renting one when needed for a short period is closely tied to the growth of homogeneous residential areas distant from work and shopping, of commuting long distances on a Monday-through-Friday schedule, of the construction of high-speed limited-access highways. These arrangements were encouraged by local and national government policies in the postwar years of road building, zoning against mixed-use neighborhoods, tax deductions for interest paid on home mortgages, and in the US by encouraging the collapse rather than revitalization of existing mass transit and railroad systems. There was too little zoning against sprawl development and a failure to maintain boundaries between dense settlements and surrounding green space. These policies could be reversed, though it would take years to see alleviating effects.

The five-day work week, hard won by labor unions, is nearly obsolete in the sense that many office chores can be handled from home via internet or telephone, and that business contacts are increasingly in other time zones, making a local 9-to-5 workday too limited. With women almost fully in the workforce, daily travel to a job may conflict nearly intolerably with the need for child caregivers at home. It would take little change in infrastructure to allow working days from home rather than office, reducing the number of commuting trips, and flexible hours to avoid rush hour commuting.

Proposals to reduce automobile use usually focus on monetary incentives, regulation, or changing technology. Higher taxes on gasoline and on the sale of fuel-inefficient vehicles would likely reduce gas consumption (in a way most punitive to those with lower income), but taxes are unpopular. When gasoline prices peak, which is functionally the same as adding a higher tax, politicians often urge that prices at the pump be lowered by one means or another.

During the oil shocks of the 1970s, a time when American cars were large, heavy gas guzzlers, the US government seriously attempted to improve the energy efficiency of cars and light trucks through regulation. The Corporate Average Fuel Economy (CAFE) standard, enacted by Congress in 1975 and put into effect in 1978, set limits on the fuel efficiency of vehicles made for sale in the US. CAFE had stricter standards for passenger cars than "light trucks," despite the majority of light trucks (including vans and SUVs) being used as passenger cars. In 1979 the standards, averaged over all vehicles a manufacturer made for sale that year, was 19.0 mpg for passenger cars and 15.8 mpg for four-wheel-drive light trucks. An automaker must pay a penalty if its annual fleet mileage falls below the standard. Fleet average is used to allow people who wish and can afford to use highly gas consuming cars to continue to do so, as long as these are offset by others using fuel efficient cars.

In the first years after the CAFE standards were put in place, fuel economy rose dramatically, abetted by the rapidly rising price of gasoline. Subsequent improvement in gas efficiency stalled and then worsened regardless of whether gas prices were rising or falling. This was coincident with the rising popularity of SUVs and other light trucks, more profitable than cars for automakers, their market share growing from 10 percent in 1979 to peak at 50 percent around 2000. At that time vehicle fleets of the US and Canada had the poorest overall fuel economy among industrial nations, while the American traffic fatality rate was worse than in other developed nations.

In 2007 the CAFE standard was strengthened despite resistance from car manufacturers, with a goal of 35 mpg by 2020 for cars and light trucks. In 2011, while distressed American automakers were recovering from near financial collapse with the help of federal bailouts, they agreed to a dramatic increase of fuel economy to 54.5 mpg by 2025. If accomplished, that is, if the industry or Congress does not renege, this would put American fuel standards on a par with Europe and Japan.

Before the energy crisis, when the cost of gasoline in America was trivial, Detroit had no concern about gas economy. Its cars were large, heavy, and powerful, weighing ten to 20 times as much as the driver. Decorated with fins and chrome, the exterior designs changing from year to year, they were expected to be replaced within about 50,000 miles. It was not unusual for drivers to buy a new model each year.

The first post-CAFE improvements in efficiency were affected mostly by reducing size and weight of cars, but this made passengers more vulnerable to injury in case of collision with a larger vehicle or fixed obstacles, perhaps causing one or two thousand additional fatalities per year. By the mid-2000s, the increasing market share of SUVs and other light trucks brought average vehicle weight back where it was before the CAFE standards were enacted. In the meantime, smaller cars became safer, largely because of improved shock absorption on impact. Light trucks were implicated in more accidents, especially rollovers because of their high center of gravity (Insurance Institute for Highway Safety 2006). This largely eliminated the tradeoff of safety for fuel economy. Also, laws enacted over the objections of automakers required life-saving seatbelts and airbags. In the future, small aerodynamic cars made of very strong but light carbon fiber may further improve safety and fuel economy simultaneously (Lovins 2011).

The past decade has demonstrated the skill of automotive engineers in safely improving efficiency once manufacturers were pressed to do so by government regulation and by buyers hit with steeply increasing gasoline prices. Japan's Toyota Prius hybrid was a leap forward, its 2010 hatchback sedan getting 50 mpg. Prior cars used only a few percent of the energy in their gasoline to move the vehicle forward. Nearly all the rest was wasted as heat thrown into the environment, whether from friction within the drive train, or the heating of brake linings when a driver stopped the car, or in idling while waiting for a green light. Toyota's clever idea was to apply some of that wasted energy to driving the car, doubling or tripling the percentage of energy in gasoline that actually moves it ahead.

A hybrid electric vehicle (HEV) like the Prius uses regular gasoline to fuel a conventional internal combustion engine, but it also has an electric engine. Both engines are capable of driving the car forward, working either by themselves or in tandem, as required. The electric engine is powerful enough to handle low-performance engine requirements on its own. When the car is idling or accelerating at low speeds, gasoline is not consumed. When additional power is required, the gas engine kicks in.

The electric engine is powered from onboard batteries that are continually recharged with energy that is wasted in conventional cars. One device for this purpose, *regenerative braking*, has been used on railroads for decades. In conventional braking, the excess kinetic energy of the vehicle is converted to heat by friction in the brake linings and therefore wasted. In regenerative braking the vehicle slows by putting its kinetic energy to work turning an electrical generator. Engaging the generator provides the braking effect. Thus, each time the driver brakes the car, she is charging the batteries. The gasoline engine also charges the batteries when the car does not need its full torque as when coasting downhill.

Numerous HEVs are now on the market, and some models are pushing farther toward the electric side. Several automakers offer "plug-in" hybrids, their batteries chargeable by plugging the car into an electric wall socket. The plug-in Chevy Volt runs entirely with an electric motor, carrying gasoline only if needed en route to run a generator to recharge its batteries. Electric motors are light, efficient, and easily change speed, reducing the need for heavy transmissions.

Unfortunately, battery technology is not sufficiently advanced to offer lightweight batteries that hold sufficient charge to run a car at adequate power for a long distance. The range of a Chevy Volt, without gasoline supplementation, is less than 40 miles. Batteries are bulky, taking up considerable cargo space, and while reliable for eight or more years of use, they are expensive to replace and require safe disposal. Also the technology presently requires rare earth elements of limited availability.

A fully electric car runs efficiently and quietly, with no tailpipe emission of CO_2 or other pollution. Until better batteries are available, relatively frequent charging would be handled during long trips with an infrastructure of service stations where batteries might be quickly switched or recharged. The central problem is that the electricity must come from somewhere – a point often ignored in popular imagination. Changing a substantial portion of a nation's vehicles from gasoline to electricity would put enormous demands on the power grid. (Recall that the typical thermal power plant operates at one-third efficiency, so two-thirds of the energy used to fuel the plant would be lost before the electricity reaches the car.) If electricity for the cars were generated by burning coal, as is presently true for nearly half of American electricity, then new coal-fueled power plants would themselves emit considerable CO_2 and other pollutants.

If a fleet of electric cars were charged largely by coal-burning power plants, would the overall amount of CO_2 produced be more or less than from a fleet of gasoline-burning cars? According to the Union of Concerned Scientists, if an all-electric car like the Chevy Volt received all its electricity from coal, then its addition to atmospheric CO_2 would be about the same as a gasoline-powered car getting 30 mpg (Anair & Mahmassani 2012). Since most cars on the road today get poorer fuel economy than 30 mpg, replacing them with electric cars would be an improvement. To the extent that electric cars were charged by non-carbon (nuclear or renewable) power plants, the improvement would be proportionately better. On the other hand, gasoline-burning HEVs like the Toyota Prius with its 50 mpg would probably produce about the same or less CO_2 than electric cars under most realistic conditions.

Some enthusiasts propose supplying electric vehicles entirely with renewable or at least carbon-free electricity, thus avoiding any increase in greenhouse gases. But in that case, one might as well use the new non-carbon generators to directly replace existing coal plants, which would reduce greenhouse gases without need for electric cars.

Another approach is to replace cars that burn gasoline with cars fueled by hydrogen (usually powering a fuel cell). The chemical equation for oxidation of hydrogen is $H_2 + O_2 \rightarrow H_2O$ + heat. Thus, the only tailpipe output from a hydrogen-powered car is water vapor. Otherwise it is as emission-free as an electric car. A hydrogen car need not be limited in range, though it does require a high-pressure tank to hold cold liquefied fuel, which is gaseous at room temperature. The central problem is where to get the hydrogen. The element is plentiful, but being highly reactive, it occurs naturally in combination with other elements and would have to be separated from some feedstock like seawater. Water is easily decomposed into hydrogen and oxygen by the process of electrolysis, a routine topic in every introductory chemistry course. (An electric current is passed through a beaker of salt water, producing bubbles of hydrogen at the negative electrode and bubbles of oxygen at the positive electrode.) But electrolysis of seawater requires too much electricity to provide a practical supply of hydrogen fuel. Natural gas (methane) is a more plausible feedstock, using less electricity to decompose: $CH_4 + O_2 \rightarrow CO + H_2 + H_2O$. This provides hydrogen fuel, but another product of the reaction, CO, quickly oxidizes to CO_2. Thus, while the hydrogen car itself does not produce carbon emissions, the facility that produces the hydrogen does. Overall, there is no reduction in CO_2.

The Jevons paradox

Any attempt to reduce energy consumption by increasing efficiency runs into the problem called "rebound" or "the Jevons paradox," named for the English economist William Stanley Jevons, who noted in 1865 that increased efficiency of coal in steam engines did not reduce demand for coal but, to the contrary, led to its increased consumption in a wide range of industries. The paradox applied to cookies would work like this: An overweight person eats a cookie a day containing 200 calories. Seeing advertised a "lo-fat" brand with only 100 calories, he switches to that cookie but now, self-satisfied, eats three per day. No doubt such retrogression sometimes occurs, but many experts doubt that the saving of energy from increased efficiency is greatly contravened by increased usage (Greening et al. 2000, but see Sorrell 2009 and Owen 2011 for contrary views).

Consider efforts to reduce gasoline consumption by requiring increased efficiency from passenger vehicles. The National Academy of Sciences concluded that in the absence of CAFE standards, motor vehicle fuel consumption in the US would have been approximately 14 percent higher than it actually was in 2002 (NAS 2002). Opponents to government regulation disagree, arguing that increased efficiency adds to the initial cost of a car, so older, less efficient cars remain on the road longer. Also, getting better fuel economy may induce some drivers to travel farther or more often than they otherwise would (the Jevons paradox). With over three decades of perspective since CAFE was put in place, we can see what actually happened. The outcome is less clear than ideological positions assert.

Figure 13.1 shows gallons of gasoline (including ethanol and other additives) produced for sale in the US per month per capita. Per capita consumption was rising in the early 1970s until the OPEC oil embargo of 1973. This caused a rise in gasoline prices and a short-lived downturn in per capita gas consumption. Gas prices generally moderated after this first oil shock, and per capita consumption resumed its upward climb, reaching the highest level ever in 1978. But in 1979 gas prices rose steeply again with a second oil shock caused by the Iranian Revolution. CAFE went into effect in 1978, nearly the same time. The reduction in per capita gas consumption from 1978 to about 1982 is usually attributed by analysts to both higher prices at the pump and the CAFE requirements; it is impossible to separate one cause from the other.

By the mid-1980s gas prices had stabilized, and automakers (with customer collusion) essentially made an end run around the CAFE standards with their newly popular and highly profitable SUVs, which were then classified as "light trucks." (At that time light trucks were under less stringent fuel economy requirements than automobiles.) Until the end of the century, per capita gas consumption actually increased though never approaching the level of 1978. Only in the past decade has per capita gas consumption improved again, and this may be explained by the rise in gasoline prices during the 2000s.

Since automakers and consumers were in effect negating the intent of the CAFE standard by replacing (or supplementing) autos with SUVs and other light trucks,

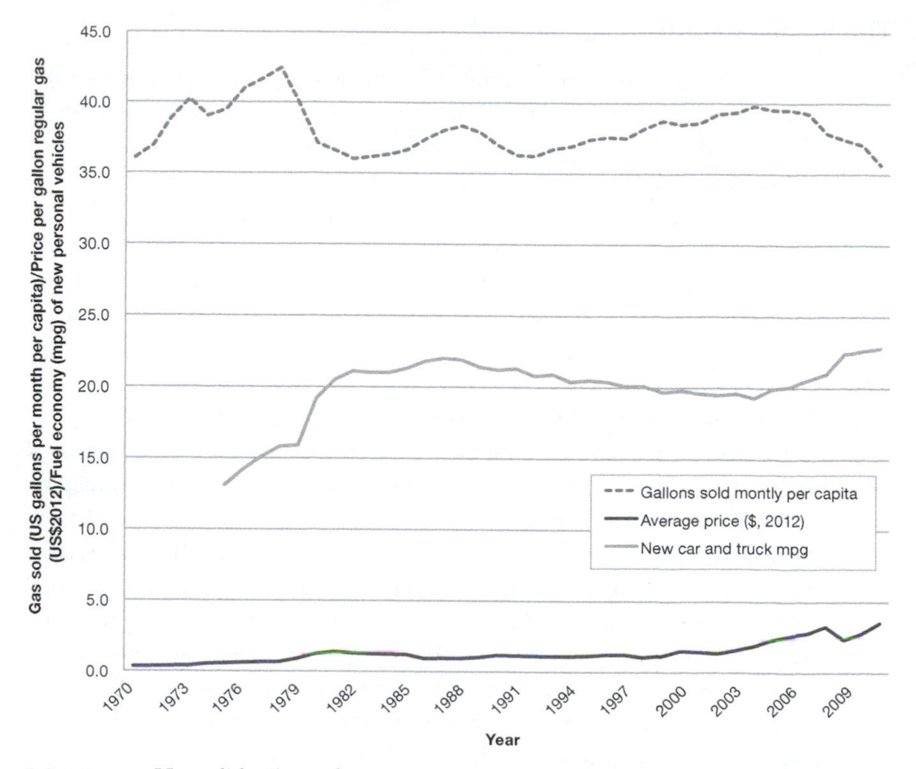

FIGURE 13.1 How did US gasoline consumption change with changes in gas prices and fuel economy of new passenger vehicles? (EIA, EPA)

the overall fuel efficiency of each year's newly sold fleet of passenger vehicles (cars plus light trucks) improved very little after 1980. This is shown in Figure 13.1 by EPA estimates of fuel economy for *all* new passenger vehicles, including not only cars but SUVs and other light trucks (EPA 2012: Table 2). Only in the past decade, as gasoline prices have risen, has overall fuel efficiency improved.

What can we conclude from this picture? First, Americans continue to consume a lot of gasoline per capita, not as much as in 1978 but barely less than in 1982. Second, gas consumption is responsive to price rises in the short term but appears to rebound after a few years of adjustment to the higher price. Third, CAFE standards were not very effective in improving the nation's overall fuel efficiency of passenger vehicles because they contained an escape clause that allowed (if not encouraged) the growth of gas-consuming SUVs and other light trucks for passenger use. Fourth, the worst prediction of the Jevons paradox, greatly *increasing* gasoline consumption, did not occur. Fifth, overall fuel economy has improved in the past decade, as it did from 1978 to the mid-1980s, though again, we cannot say if this is due to CAFE regulation of higher gas prices.

The lesson here is that the Jevons paradox is too simple to apply as a fixed rule. The degree of rebound, if any, from efficiency regulations or from more efficient technologies will depend on the specifics of each case. If there is profit to be made by skirting efficiency standards, then we should expect that to occur. Some efficiency regulations, including the initial CAFE law, seem to invite skirting. If that was not anticipated at the outset, it was clearly apparent by 1990 when the law could have been amended, but it was equally apparent that the Congress and the automobile and oil industries had no real interest in improving fuel efficiency.

Efficient buildings

About two-thirds of the large, privately owned buildings in New York City are multifamily properties, another quarter is office buildings, and most of the rest are retail stores. The variability in energy efficiency among buildings in the same category is remarkable. Among multifamily and office buildings, the most efficient five percent use less than a quarter of the energy consumed by the least efficient five percent (for the same amount of floor space). Among large retail stores, the most efficient use little more than a tenth as much energy as the least efficient (again for the same floor space). If all the comparatively inefficient large buildings were brought up to the median energy efficiency in their category, New York City could reduce energy consumption in large buildings by roughly 18 percent (New York City 2012).

Buildings, including the appliances in them, use more than 40 percent of America's energy and nearly three-quarters of its electricity, much of it wasted. In recent years, US buildings used more primary energy than the total energy used by Japan or Russia, and twice that of India's 1.2 billion people. If American buildings were a nation, they would rank third, after China and the United States, in primary energy use (Lovins 2011). The building sector offers an enormous opportunity to cut back on valueless energy and electricity consumption.

Energy Star is a program created in 1992 by the US EPA and the Department of Energy to set voluntary energy-efficiency standards for consumer products including computers, kitchen appliances, furnaces, air conditioners, and water heaters. Subsequently adopted by the other industrial nations, the Energy Star label on appliances increases their popularity with consumers, indicating that a device uses 20 to 30 percent less energy than required by federal standards. Residential and commercial buildings are also eligible for Energy Star ratings.

Recent building codes in many areas of the United States impose good insulation standards on new construction and appliances, but could go further in discouraging 5,000 square foot (465 m^2) homes with attached three-car garages for a family of four, perversely advertised as "energy efficient." Older leaky or uninsulated houses, common in poor neighborhoods, are especially wasteful of heating fuel during winter. Cost-effective weatherization not only saves energy and lowers utility bills but makes interior spaces more comfortable and creates local jobs for semi-skilled workers who do the retrofitting. Like today's furnaces, water heaters, and lighting that are far more efficient than what they replace, weatherization can produce enough savings on utility bills that over five to 15 years it pays back the initial investment. On the other hand, there are numerous modern devices – flat-screen TVs, microwaves, computers, and phone chargers – that consume standby power even when not being used or on "sleep" mode, which may cause up to 10 percent of residential electricity use. Microwave ovens probably use more electricity for their illuminated clock than for cooking (Lovins 2011). These could be switched totally off when not being used. Programmable thermostats save heating and cooling costs without inconvenience. Motion-sensing switches turn off lights when no one is in the room.

These are all easy fixes, the technology is available, but social and psychological impediments keep many property owners from implementing them. Often owners lack information, interest, or motivation, and behavioral inertia (or laziness) sustains wasteful habits. Potential savings on utility bills, if recognized, seem too small or distant in time to warrant a large upfront investment in energy improvements. Natural gas, commonly used for space and water heating, is currently inexpensive, reducing the incentive for new furnaces and water heaters. Even when incentives greatly shorten the payback period, many homeowners and businesses resist spending money on capital improvements because they do not expect to stay in that place very long. Landlords have no incentive to reduce utility bills paid by their tenants.

The same barriers prevent most home and building owners from installing solar water heaters or photovoltaic panels on their roofs even when they would have relatively short payback periods. The sun-drenched cities of Nevada have surprisingly few rooftop panels. In the state hardest hit by the financial crisis of 2008, few Nevadans have enough equity in their homes to finance solar systems. The state's investor-owned utility company, NV Energy, does meet a substantial renewable portfolio requirement, primarily through geothermal and centralized solar facilities. Rather than going beyond the requirement by exploiting the dispersed solar potential of the state's rooftops, NV Energy and its customers prefer the cheap electricity generated by the utility's old coal-fired power plants that environmentalists contend should have been retired years ago.

In Nevada and elsewhere, what is needed to promote energy improvements in existing individually owned buildings, whether weatherization or rooftop solar systems, is a simple means for the owner to apply for a cost-effective improvement, to have it installed without any upfront payment, and to receive immediate savings on monthly utility bills. Such programs would have to be run by financially powerful agents, whether federal or local governments, utility companies, or private equity firms. The agent would wholly finance energy improvements to qualifying buildings, recouping its investment plus interest from the monthly saving on each building's utility bill. A portion of the saving would go to the owners, fulfilling a guarantee that they would see an immediate reduction in monthly utility payments.

A few American communities and companies are experimenting with financial systems more or less on this pattern, allowing retrofits to existing homes and monthly savings without any upfront costs (Himmelman 2012).

Mandatory building codes for energy efficiency are controversial in the United States but commonplace in the European Union where all new or renovated buildings, and existing buildings at the time of sale, must receive energy performance certificates. "Smart meters" displaying electricity use must be installed in new or refurbished buildings. Those constructed after 2019 must generate as much energy as they use. Some European cities have more stringent energy laws than those of the EU. Barcelona, taking advantage of its sunny climate, requires that solar energy supply 60 percent of hot water in all new or renovated buildings.

US energy efficiency regulations for buildings are more modest than in Europe and usually voluntary. As of 2009, 92 cities had laws requiring large new buildings to meet some kind of minimum standard. The U.S. Green Building Council is a nonprofit trade organization that in 1998 created its Leadership in Energy and Environmental Design (LEED) rating for new buildings. This is a voluntary certification that can lower long-term operating costs and is popular with developers for its cachet in selling or leasing new buildings for higher prices. LEED has raised awareness among American architects and builders of "green building" design and construction. David Owen (2009) suggests that it is now virtually unthinkable to erect a major building without at least announcing some intention to seek LEED certification.

Among the most common criticisms of LEED certification is that the process is complicated, costly, and time consuming; that energy savings are modest because that is just one of five areas in which buildings can earn a high rating; that certification pays more attention to hi-tech add-ons than to simpler measures for conserving energy; and actual energy performance need not be monitored once a building has been certified. Indeed, there is some indication that LEED buildings use slightly more energy than similar non-LEED buildings of the same age (Owen 2009; Troy 2012).

Cogeneration

Recall that the typical thermal power plant generates electricity at one-third efficiency, so three energy units of primary fuel must be burned to produce one energy unit of electricity. This occurs because the steam that turns the turbine must be cooled in the plant's condenser in order to do its work; thermodynamic limits preclude much improvement in efficiency. Heat coming from the condenser is released to the environment through cooling towers, nearby bodies of water, or flue gases. Since the United States uses nearly half its total primary energy supply to generate electricity, it "throws away" nearly a third of TPES in the process, an enormous waste of energy.

Cogeneration, also called "combined heat and power" (CHP), puts this otherwise wasted heat to good use, potentially utilizing 80 percent (rather than 33 percent) of the energy in the primary fuel. Usually this is accomplished by piping steam or hot water from the power plant's condenser through underground pipes to nearby buildings for space or water heating, or as a heat source for industrial processes. A car engine becomes a CHP in winter when some of the reject heat is used to warm the vehicle interior. CHP is one of the most cost-effective means of reducing carbon emission in wintertime or in cold climates (http://en.wikipedia.org/wiki/Cogeneration).

The typical single-family home contains its own furnace and water heater, their fuel costs tallied on monthly utility and water bills. There are efficiencies in using one central steam plant for a cluster of buildings, as often occurs on university campuses. Entire neighborhoods or cities could be heated by central steam plants, a system called "district heating." Cogeneration is ideally suited to district heating, with one or a few electric generating plants serving as the central heat sources for most of a city's buildings. This rarely occurs in the United States though an exception already mentioned is New York City, where Consolidated Edison sends steam at 180° C (350° F) from the condensers of its electrical generators to 100,000 buildings in lower Manhattan, one factor that makes the area energy efficient. It is the biggest cogeneration district in America. CHP is most efficient when heat can be used close to the generation site, as in nearby apartment buildings or factories. More distant or dispersed uses require heavily insulated pipes and raise the cost.

The European Union actively promotes CHP in its energy policy and generates 11 percent of its electricity using cogeneration, though use varies widely across member nations. In Denmark about half of space and water heating is provided by cogeneration, in Finland about 40 percent. In most Russian cities, district level CHP plants provide about half the nation's electricity and send hot water to neighboring city blocks. Germany and the UK plan considerable increases in CHP. Recently the US Department of Energy established eight research centers to promote CHP, waste heat recovery, and district energy technologies; however, outside of Manhattan the opportunity is not much exploited.

Cogeneration can be used in individual buildings, though probably at lower efficiencies than centralized systems. Japan's Honda Motor Company sells compact residential-use cogeneration units ("micro turbines") that run on natural gas and generate electricity for home use while utilizing the exhaust heat to provide hot water (http://world.honda.com/powerproducts-technology/cogeneration/).

Conclusion

Solving our energy problems is not a task for the simple minded. Promising directions are apparent, but all have pitfalls, and none has a political consensus. Barriers are not technological as much as sociological. The industrial nations do not need to use the amount of energy and electricity that they presently consume, and are on course for consuming in the future. They waste so much energy, none more than the United States, that it is feasible to cut far back without sacrificing quality of life. But there are too many special interests, ideologues, ignoramuses, and seriously concerned people who do not agree on what to do. Still, solutions have been tried, more so in Europe and Japan than in North America, with some successes and some failures. More experience and analysis can clarify course corrections in case governments ever agree to take action.

14

WHAT NEXT?

The most highly consuming industrial nations use far more energy and electricity per capita than they need to maintain their comfortable lifestyles. They have major social institutions – corporations, government agencies, and professional organizations – that exist to feed and expand our appetite for fuels and electricity. These bring abundant benefits at relatively low monetary cost, spurring consumer demand, but our increasing consumption of fuels, whether finite or renewable, will inevitably worsen an array of problems. Technology can make more energy available and improve the efficiency with which we use it. But unless we modify the corporations, government agencies, and professional organizations that drive increasing energy/electricity use, and the patterns of business and consumer behavior that evolved to use "cheap" and abundant energy, those problems will intensify.

There never was a Garden of Eden or any other place where human life was more than momentarily better than it is for most citizens of today's industrial countries, notwithstanding our recent battering by financial crises. It has been only a century since we began to realistically expect all our babies to survive us, reaching old age with a material lifestyle, human rights, and breadth of opportunity not enjoyed even by aristocrats in Victorian England. Anyone who believes there was once a time and place when things were better should read a good history book about that time and place.

There has always been and always will be human tragedy, or threat of tragedy, on an epoch scale. Today there is no discernible threat to the industrial world that to my mind equals World War II or, far worse, the nuclear arsenals of the United States and the Soviet Union during the Cold War, which if triggered would have detonated 10,000 hydrogen bombs over each nation, obliterating life in the Northern Hemisphere. The most extreme plausible scenario for global warming does not approach that level of calamity. In my own view, the worst we are presently doing to the earth is keeping so many people of the Third World in poverty, illness, and unhappiness; and bringing widespread extinction to nonhuman species at a rate unprecedented in human history.

If we solved our energy problems, non-energy problems as bad or worse would remain. There will be pandemics, wars, social inequities, congested cities, financial crises, clean water shortages, industrial catastrophes, and no end to natural disasters causing hundreds of thousands of deaths. Use your imagination to continue the list. The world will never be a bed of roses. My best hopes are that each of us and our families live a fulfilling life, and that each generation leaves this place better than they found it, not necessarily more natural and less manufactured, but more cultured and learned, more humane, more livable for species beside ourselves, more equitable and just.

The foregoing paragraphs read more like a manifesto than a prescription for change, but it is essential to cut through excessively gloomy images of the present and utopian fantasies of the future. These as well as self-serving politics, crass profit seeking, and willful scientific ignorance prevent even so technically advanced a nation as the United States from adopting a progressive and pragmatic energy policy, one based on the best knowledge that science, engineering, and sociology can offer.

What do we know?

A quick overview of prior chapters: High consumption of hydrocarbon fuels causes climate change and other serious problems. The challenge is to cut their use, not because we are imminently running out of exploitable deposits, but because fuel usage in such volume has so many undesirable consequences. Cutting will be difficult because hydrocarbons are strongly favored by corporate interests, existing infrastructure, and because of their relatively low monetary costs. Looking to the Third World, there is real need for cheap and plentiful energy in China, India, and other agrarian nations aspiring to improve their living conditions.

Every energy source has advantages and disadvantages. Proponents of each source tend to exaggerate its benefits and minimize its harms; opponents do the opposite. Partisans may reach their positions through rational calculation, emotional reaction, or ideology; they may or may not stand to profit from using one source rather than another. In any case, no energy policy, however wisely drawn, will satisfy everyone.

No non-carbon source of energy is trouble free, inexpensive, or without public opposition. There is virtually no prospect that most industrial nations will completely or even mostly move to non-carbon sources in the near future and certainly not China or India.

We have seen that when per capita energy/electricity consumption reaches the minimum level for industrial countries, there is little if any measurable improvement in quality of life from consuming even more. Growing population, the primary Malthusian worry, does place some added burden on total primary energy supply.

But since World War II, increases in electricity consumption, promoted by energy companies, far outpaced the growing number of people.

In the United States, nearly half of all primary fuels are used to produce electricity. That portion is not as large in other nations, but it is growing nearly everywhere. Since the typical thermal electrical generator requires three energy units input as fuel for each energy unit output as electricity, it is as if two-thirds of the energy devoted to electricity production were thrown away (as waste heat) before any electricity reaches the end user. The efficiency of converting gasoline's chemical energy to kinetic energy of a moving automobile (with a conventional engine) is even worse. Consider too that the engine typically pushes a heavy car to transport a far lighter person, often traveling farther and more frequently than is really needed or desired. There must be better, less wasteful ways to achieve comparable ends.

This excessive consumption and misuse of energy is for sociological as well as technological and financial reasons. Its ultimate cause is the industrial transformation that brought great benefits to Europe, North America, and Japan while making these areas reliant on huge quantities of fossil fuels. It is pointless to ask if the history of industrialization might have been different; what's done is done. But China and other nations now trying to modernize should inquire if they must travel the same path, suffering the same pitfalls. Must they consume so much energy and electricity, especially by burning fossil fuel, to achieve a satisfying lifestyle? Are there better ways to organize transportation, housing, work, and leisure than the high-consumption patterns that developed in the West? Can we dissipate the momentum of ever onward energy interests?

What should we do?

This is not the place to plan a specific route to our energy future, but it is worthwhile raising some signposts to help point the way:

1. Don't wholly foreclose any plausible source of energy. A source that seems good (or bad) today may look worse (or better) tomorrow. In the 1970s, when the hazards of coal mining and recognized pollutants from burning were mitigated, coal seemed an excellent fuel for producing electricity: cheap, abundant, and energy intensive. By the 1990s, after global warming emerged as a severe problem, coal, the worst emitter of CO_2, became a pariah. If the industry ever develops a practical means of capturing CO_2 emissions, coal will again be a fuel of choice.
2. Renewables have undeniable virtues but are not ipso facto preferable to finite fuel. A renewable source usually costs more money, and when its entire fuel cycle is considered, may be more harmful to health or damaging to the environment than a finite source carefully used. Corn ethanol, though renewable, may be a worse option than natural gas.
3. Monetary cost (or profit), usually calculated over the short term, is the primary motivator for choosing what energy sources are exploited, which technologies

are developed for production and transport of fuels, and the amounts of energy and electricity that are produced and consumed. Enthusiasts for a costly new or underused energy source often argue that after a period of subsidized development and use, its unsubsidized price will fall to that of traditional fuels. That may be true, but the forecast ignores the likely reaction of coal and oil companies and fuel-rich nations that have access to decades if not centuries of exploitable fossil fuels. Faced with a loss of market share, they will lower their own prices. At first blush this seems a desirable outcome: cheap energy. But lower prices are almost guaranteed to increase consumption, thus worsening all the other problems of energy overuse.

When market-driven choices bring societally bad consequences such as pollution, climate change, illness, resource wars, or destruction of habitat and landscape, then government has the task of disallowing certain choices through regulation. Alternatively, government may alter costs (and profits) through subsidies, tax policy, grants, and penalties. Leaving energy choices wholly in the hands of corporate or entrepreneurial interests is foolish.

4. It is appalling (but unavoidable in nations committed to free speech) that partisans driven by political, financial, or ideological interests purposively deny or distort solid scientific findings in order to advance their own agendas. Strenuous effort to discredit the authoritative and transparent IPCC reports on climate change is the most notorious but not the only instance of self-serving obfuscation.

   There is a high degree of scientific illiteracy even in as technically advanced a nation as the United States, where half the population does not believe that humans evolved from other primates (Mazur 2008). Truth, for most adults, is what they are told by their family and friends, their church, their political party, or their preferred news media. Even the best news organs have been criticized – sometimes deservedly, sometimes not – for inaccurate and biased reporting about technological risks. A related problem is the uncritical acceptance by some journalists of scientific claims made by biased or incompetent sources (Singer & Endreny 1994; Mazur 1998; Lichter & Rothman 1999).

   In democracies the press guards its own practices, sometimes rigorously. A peculiar instance was the World War II agreement among American journalists to avoid showing President Franklin Roosevelt as a cripple. Today the press self-enforces bans on language and visuals that violate common standards of decency; on insults to minority groups; and on identification of confidential sources even when such a ban denies accused parties knowledge of their accusers. Journalists exercise admirable responsibility in such matters. Perhaps they could add to that list some self-restraint on uncritically reporting overtly insubstantial scientific claims.

5. Any plan for a new energy facility or electrical transmission system should consider the likelihood and impact of public opposition. Even authoritarian China is now coping with previously unheard-of public protest over

development projects. In industrial democracies, we should expect that any major new project will engender controversy, and if given sufficient attention by news media may become mobilized opposition. This is not in itself bad. Opponents sometimes discover or highlight serious deficiencies in engineering proposals and may deter the profusion of deleterious projects.

6. Industrial nations do not use energy efficiently. As much attention should be given to increasing efficiency as to seeking new sources of energy. Cogeneration, hybrid cars, and other schemes that put "waste heat" to use are especially promising.
7. Major social institutions – corporations, government agencies, professional associations – exist to enlarge the production and consumption of energy and electricity. Any long-term plan that does not address and diminish the momentum of these institutions will not affect sustained conservation.
8. Increasing dependence on a single, highly integrated power grid to serve an entire nation or multinational region raises the possibility of very widespread blackouts. Some dispersion into separate grids or distributed generators is desirable to avoid system-wide failures.
9. For the world's richest people – especially Americans, Canadians, and Australians – energy is too cheap, encouraging waste or at least discouraging better efficiency. If improving efficiency reduces energy prices further, consumption will likely increase to some extent (the Jevons paradox). Improved efficiency will not be optimally effective without doing something to limit overall energy consumption.
10. China, India and other Third World nations emulate the material habits of industrial nations, most particular the highly consuming, energy-crazed United States. (Denmark would be a better model.) They buy, license, or copy industrial technology for energy production, electricity generation, and extensive power grids. Third World consumers adopt our energy-intense desires for personal automobiles, large homes, and all the modern appliances. We should be leading these aspirants in a direction that would improve rather than worsen future global problems.

Appendix I

CHEMISTRY OF AIR POLLUTION

For readers wishing to better understand air pollution, it is worth delving into some high school chemistry: Molecules are identified by the elements from which they are composed, for example, a molecule of water is H_2O, where the subscript 2 indicates that two hydrogen (H) atoms are bound to an oxygen (O) atom. A molecule of carbon dioxide contains one carbon atom and two oxygen atoms, thus CO_2. Earth's atmosphere is 78 percent nitrogen and 21 percent oxygen. If you plucked an oxygen molecule from the air, you would normally find not a single atom but two oxygen atoms bound together, denoted O_2. The same is true of the nitrogen gas in the atmosphere, denoted N_2.

A chemical reaction, when the atoms in two or more kinds of molecule rearrange to form different molecules, is described by a chemical equation. On the left side of the equation are the molecules present before the reaction occurs, and on the right side are the molecules produced by the reaction. Chemists use arrows rather than equal signs to indicate the direction of the reaction. The simplest fossil fuel is methane (the major component natural gas), one carbon atom bound to four hydrogen atoms, denoted CH_4. Complete oxidation (burning) of methane is simply expressed:

$$CH_4 + 2O_2 \rightarrow CO_2 + 2H_2O + \text{heat.}$$

The equation says in shorthand that by combining a methane molecule with two oxygen molecules (i.e., by oxidizing, or burning, methane), you get a molecule of carbon dioxide plus two molecules of water plus the release of heat. The reason that heat is produced by oxidation is that there was more energy stored in the chemical bonds of the CH_4 and O_2 on the left side of the equation, than in the chemical bonds of the CO_2 and H_2O on the right, so that excess energy is released as heat, which of course is the reason we burn fuel.

If the combustion is not complete, then the pollutant carbon monoxide (CO) is also produced in the reaction. People used to commit suicide by running a car engine in a closed garage and "going to sleep" from inhaling the CO. Carbon monoxide has been greatly reduced by ensuring more complete oxidation of gasoline with catalytic converters in vehicle exhaust systems. I sometimes hear that you cannot kill yourself anymore with car exhaust but would not test the claim.

Atoms are not destroyed or created during a chemical reaction, only rearranged into different molecules. The number of each kind of atom on the left of the equation must equal the numbers on the right. To make the equation "balance," i.e., to keep the number of each type of atom the same on both sides, we need two oxygen molecules for each methane molecule, and we get two water molecules for each carbon dioxide molecule. That is what happens in nature. Balancing becomes complicated for petroleum, whose molecules are diverse with large and varying numbers of Cs and Hs, so to simplify, I will write any hydrocarbon molecule as *HC*, where italic means that the numbers of hydrogen and carbon atoms in the molecule are unspecified, and I will not bother balancing subsequent equations. The equation for oxidation of methane, or any other hydrocarbon, then becomes simpler:

$$HC + O_2 \rightarrow CO_2 + H_2O + \text{heat}$$

Beside H and C, coal and petroleum often contain other elements including sulfur (S), an especially troublesome presence. Then oxidation looks like this:

$$HC + S + O_2 \rightarrow CO_2 + H_2O + SO_2 + \text{heat}$$

SO_2 (sulfur dioxide) is a health hazard, and it combines with water vapor (H_2O) in the atmosphere to produce H_2SO_4 or sulfuric acid, a major component of acid rain. It can react with other compounds in the air to form particulates that are inhaled, causing or worsening respiratory diseases such as emphysema and bronchitis, and can aggravate existing heart disease, leading to increased hospital admissions and premature death.

At normal temperatures the molecules of nitrogen (N_2) and oxygen (O_2), abundant in the atmosphere, do not react with each other, but at the high temperatures of fuel combustion, nitrogen also oxidizes, making the full fuel-burning reaction:

$$HC + S + O_2 + N_2 \rightarrow CO_2 + H_2O + SO_2 + NO + NO_2 + \text{heat}$$

NO and NO_2, the two molecular combinations of nitrogen and oxygen, are together denoted NO_x (pronounced *nox*, rhymes with rocks). These are the oxides of nitrogen that are limited by the Clean Air Act. NO_2 is a respiratory irritant, especially in people with asthma, and it gives the brown color to photochemical smog.

Like SO_2, NO_x combines with water vapor (H_2O) to form an acid (HNO_3, nitric acid), a component of acid rain that is corrosive, damaging buildings and contributing to the acidification of soils and lakes, where it is harmful to aquatic species. Again like SO_2, NO_x reacts with other compounds in the air to form small particles that are inhaled to morbid effect.

Photochemical smog, the most visible type of modern air pollution, was first explained in the 1950s. All cities with automobile traffic are susceptible to smog, particularly in warm dry urban areas like Los Angeles, where the situation is worsened by the city's location in a basin nearly surrounded by mountains. Because smog travels with the air, it visibly affects the agricultural Central Valley of California, even moving up into the Sierra Nevada Mountains.

Smog is produced by many complex chemical reactions between NO_x and diverse hydrocarbon molecules floating in the air because they were not completely oxidized in internal combustion engines. (These floaters are called "volatile organic compounds," or VOCs.)

One of many smog-related chemical reactions produces an unattached oxygen atom:

$$NO_2 + \text{sunlight} \rightarrow NO + O$$

This oxygen atom, desperately seeking something to bind to, might grab a normal O_2 molecule, producing O_3 (ozone):

$$O + O_2 \rightarrow O_3.$$

Or it might react with the floating hydrocarbons, forming tiny particulates, making the air opaque as well as hazardous to breathe.

Ground-level ozone in smog must not to be confused with "good" ozone high in the stratosphere, which protects living things from ultraviolet solar radiation. Ground-level ozone causes reduction in lung function and increasing respiratory symptoms, especially afflicting children, the elderly, and people with lung diseases. It is associated with increased frequency of emergency room visits, hospital admissions, and possibly premature deaths.

There are other emissions, especially from coal burning, that do not take part in the oxidation but are still troubling. Mercury and arsenic occur naturally in many rocks including coal and are released up the smokestack of a coal-burning plant. Elemental mercury is not especially dangerous, but when it reaches water, microorganisms change it into methylmercury, a neurotoxin that builds up in fish, shellfish, and animals that eat fish. Fish and shellfish are the main sources of methylmercury exposure to humans, and if eaten by pregnant women may pass into their fetuses, harming the developing nervous system.

Coal, especially lower-quality coal, contains of a lot of mineral grit, which also does not participate in the combustion reaction. Large grit particles compose the residual ash that can be toxic and in any case must be disposed of. Smaller particles, including unburned bits of impure carbon, pass up the chimney and may be removed by antipollution devices in modern plants. Nonetheless, very small particulates escape in stack gases or are produced in the air by reactions with other pollutants.

The very smallest particulates, technically known as $PM_{2.5}$ (matter with a diameter less than 2.5 micrometers, or about 1/30 the width of a human hair), often called "soot" and mostly caused by fuel burning, are now the most dangerous air pollutant in industrial nations. When inhaled they can penetrate deep into the lungs and are linked to premature death, heart attacks, strokes, worsened asthma, and possibly cancer and developmental and reproductive harm.

REFERENCES

Amin, M. (2003). *Complex Interactive Networks/Systems Initiative*. Palo Alto CA: EPRI.

Anair, D. and A. Mahmassani. (2012). *State of Charge*. Cambridge MA: Union of Concerned Scientists.

Andrews, K. and N. Carena. (2010). "Making the News: Movement Organizations, Media Attention, and the Public Agenda." *American Sociological Review* 75: 841–866.

Aoki, H. and Y. Aoki. (2010). *How to Make Mega-cities Energy Efficient*. Montreal: Annual Congress of the World Energy Council.

ANS (American Nuclear Society). (2012). *Fukushima Daiichi*. American Nuclear Society Special Committee on Fukushima.

Barboza, D. (2007). "105 Killed in China Mine Explosion." *New York Times* (December 7).

Beck, B. (1988). *Interconnections: The History of the Mid-Continent Area Power Pool*. Minneapolis MN: MAPP.

Bergen, P. (2001). *Holy War, Inc.* London: Weidenfeld & Nicholson.

Bisconti, A. (2011). *More than 70 Percent of Americans Favor Use of Nuclear Energy*. Chevy Chase MD: Bisconti Research.

Boffey, P. (1975). *The Brain Bank of America*. New York: McGraw-Hill.

BP. (2011, 2012). *Statistical Review of World Energy 2011*. London: BP.

Breyer, S. (1993). *Breaking the Vicious Circle*. Cambridge MA: Harvard University Press.

Bronson, R. (2008). *Thicker than Oil: America's Uneasy Partnership with Saudi Arabia*. New York: Oxford University Press.

Brown, M., F. Southworth and A. Sarzynski. (2008). *Shrinking the Carbon Footprint of Metropolitan America*. Washington DC: Brookings.

Burtraw, D., A. Krupnick and G. Sampson. (2012). *The True Cost of Electric Power*. Washington DC: Resources for the Future.

Busby, R. (1999). *Natural Gas in Nontechnical Language*. Tulsa, OK: PennWell.

Carpenter, M. (2012). "Dam Removal to Help Restore Spawning Grounds." *New York Times* (June 11).

Casazza, J. (1993). *The Development of Electric Power Transmission*. New York: IEEE.

Casazza, J. (2007). *Forgotten Roots: Electric Power, Profits, Democracy and a Profession*. Springfield, VA: American Education Institute.

Casazza, J. and F. Delea. (2003). *Understanding Electric Power Systems: An Overview of the Technology and the Marketplace*. New York: Wiley/IEEE.

Charles, D. (2009). "Renewables Test IQ of the Grid." *Science* 324 (10 April): 172–175.

Chase, P. (1928). "Transmission Line for the Conowingo Development." *Transactions A.I.E.E.* (July): 900–908.

Chernobyl Forum. (2006). *Chernobyl's Legacy*. Vienna, Austria: International Atomic Energy Agency.

Clarke, L. (1991). *Acceptable Risk?* Berkeley CA: University of California Press.

Clarke, L. (2001). *Mission Impossible: Using Fantasy Documents to Tame Disaster*. Chicago: University of Chicago Press.

Cohen, I. (1952). "Prejudice Against Lightning Rods." *Journal of the Franklin Institute* 253: 393–440.

Cohen, J. (1996). *How Many People Can the Earth Support?* New York: W. W. Norton.

Coll, S. (2012). *Private Empire: ExxonMobil and American Power*. New York: Penguin.

Cottrell, F. (1955). *Energy and Society*. New York: McGraw Hill.

Darmstadter, J., J. Dunkerley and J. Alterman. (1977). *How Industrial Societies Use Energy*. Baltimore MD: Johns Hopkins University Press.

Davis, C. and K. Hoffer. (2012). "Federalizing Energy? Agenda Change and the Politics of Fracking." *Policy Sciences* 45: 221–241.

Department of Energy. (1978). *Major Extra High Voltage Transmission Lines December 31, 1978*. Washington DC: Department of Energy.

Diamond, J. (2005). *Guns, Germs, and Steel*. New York: W. W. Norton.

Diamond, J. (2011). *Collapse*. New York: Penguin.

Dietz, T. and E. Rosa. (1997). "Effects of Population and Affluence on CO_2 Emissions." *Proceedings of the National Academy of Sciences* 94: 175–179.

Dietz, T., E. Rosa and R. York. (2008). "Environmentally Efficient Well-Being: Rethinking Sustainability as the Relationship between Human Well-being and Environmental Impacts." *Human Ecology Review* 16: 113–122.

Dietz, T., E. Rosa and R. York. (2010). "Human Driving Forces of Global Change: Examining Current Theories." In E. Rosa, A. Diekmann, T. Dietz and C. Jaeger (Eds.) *Threats to Sustainability: Understanding Human Footprints on the Global Environment* (pp. 83–132). Cambridge: MIT Press.

Dominy, F. (1969). "Economic Aspects of the Pacific Northwest–Southwest Intertie." *IEEE Spectrum* 6 (February): 65–71.

Edison Electric Institute. (1962). *Report on the Status of Interconnections and Pooling of Electric Utility Systems in the United States*. Washington, DC: Edison Electric Institute.

Ehrlich, P. (1968). *The Population Bomb*. New York: Ballantine.

Ehrlich, P. and J. Holdren. (1971). "Impact of Population Growth." *Science* 171: 1212–1217.

EIA. (2008). *International Energy Annual 2006*. Washington DC: Energy Information Administration.

EIA. (2011). *World Shale Gas Resources*. Washington DC: US Department of Energy.

Ellul, J. (1964). *The Technological Society*. New York: Knopf.

EPA. (2012). *Light-Duty Automotive Technology, Carbon Dioxide Emissions, and Fuel Economy Trends: 1975 Through 2011*. Washington DC: US Environmental Protection Agency.

Erol, U. and E.S.H. Yu. (1987). "On the Causal Relationship between Energy and Income for Industrialized Countries." *Journal of Energy and Development* 13, 113–122.

Farrell, A., R. Plevin, B. Turner, A. Jones, M. O'Hare and D. Kammen. (2006). "Ethanol can Contribute to Energy and Environmental Goals." *Science 311* (January): 506–508.

Federal Power Commission. (1949). *Principal Electric Facilities in the United States 1949*. Washington DC: Federal Power Commission.

Federal Power Commission. (1966). Principal Electric Facilities in the United States 1949. Washington DC.

Federal Power Commission. (1967). *Prevention of Power Failures, 3 volumes*. Washington DC.

FERC. (2000). *Investigation of Bulk Power Markets: ERCOT (Texas)*. Washington DC: Federal Energy Regulatory Commission.

Fischhoff, B. (2012). *Judgment and Decision Making*. London: Routledge.

Fleisher, J. (2008). "ERCOT's Jurisdictional Status: A Legal History and Contemporary Appraisal." *Teas Journal of Oil, Gas, and Energy Law* 3 (1): 1–21.

Fox-Penner, P. (2010). *Smart Power*. Washington DC: Island Press.

Freudenburg, W. and E. Rosa. (1984). *Public Reactions to Nuclear Power: Are there Critical Masses?* Boulder CO: Westview Press.

Friedlander, G. (1966). "The Northeast Power Failure – A Blanket of Darkness." *IEEE Spectrum* 2 (February): 54–73.

Friedman, S. (1986). *Scientists and Journalists: Reporting Science as News*. New York: Free Press.

Friedman, T. (2008). *Hot, Flat, and Crowded*. New York: Farrar, Straus and Giroux.

Fritzche, A. (1989). "The Health Risks of Energy Production." *Risk Analysis* 9: 565–577.

Gaskell, G., M. Bauer, J. Durant and N. Allum. (1999). "Worlds Apart? The Reception of Genetically Modified Foods in Europe and the US." *Science* 285 (16 July): 384–387.

Ghali, K.H., El-Sakka, M.I.T. (2004). "Energy Use and Output Growth in Canada: A Multivariate Cointegration Analysis." *Energy Economics* 26, 225–238.

Gillis, J. (2011). "Figures on Global Climate Show 2010 Tied 2005 as the Hottest Year on Record." *The New York Times* (January 12).

Gillis, J. (2012). "Clouds' Effect on Climate Change is Last Bastion for Dissenters." *The New York Times* (April 30).

Goldman, M. (2010). *Petrostate: Putin, Power, and the New Russia*. New York: Oxford University Press.

Goodell, J. (2006). *Big Coal*. Boston MA: Houghton Mifflin.

Gore, A. (2006). *An Inconvenient Truth*. New York: Rodale Books.

Graham, B. and W. Reilly. (2011). *National Commission on the BP Deepwater Horizon Oil Spill and Offshore Drilling*. Washington DC: US Government Printing Office.

Greening, L., D. Greene and C. Difiglio. (2000). "Energy Efficiency and Consumption – The Rebound Effect – A Survey." *Energy Policy* 28: 389–401.

Hachten, W. and J. Scotton. (2007). *The World News Prism*. Malden MA: Blackwell.

Hall, C. and K. Klitgaard. (2011). *Energy and the Wealth of Nations: Understanding the Biophysical Economy*. New York: Springer.

Himmelman, J. (2012). "The Secret of Solar Power." *The New York Times* (August 9).

Hirsh, R. (1989). *Technology and Transformation in the American Electric Utility Industry*. New York: Cambridge University Press.

Hirsh, R. (1999). *Power Loss: The Origins of Deregulation and Restructuring in the American Electric Utility System*. Cambridge MA: MIT Press.

Hobsbawm, E. (1969). *Industry and Empire: An Economic History of Britain Since 1750*. Baltimore MA: Penguin Books.

Holdren, J., K. Smith and G. Morris. (1979). "Energy: Calculating the Risks (II)." *Science* 204: 564–567.

Hughes, T. (1983). *Networks of Power*. Baltimore MA: Johns Hopkins University Press.

IEA. (2009). *Energy Balances of OECD Countries, 2009 Edition*. Paris: OEDC/IEA.

IEA. (2011). *Annual Oil Information 2011.* Paris: OECD/IEA.

IEA. (2012). *Golden Rules for a Golden Age of Gas.* Paris: OECD/IEA.

Inglehart, R., R. Foa, C. Peterson and C. Welzel. (2008). "Development, Freedom, and Rising Happiness." *Perspectives on Psychological Science* 3: 264–285.

Inhaber, H. (1979). "Risk with Energy from Conventional and Nonconventional Sources." *Science* 203: 718–723.

Insurance Institute for Highway Safety. (2006). "How Vehicle Weight, Driver Deaths, and Fuel Consumption Relate." Status Report 41 (February 25): 3–4. Arlington VA: Insurance Institute for Highway Safety.

Intergovernmental Panel on Climate Control. (1990, 1995, 2001, 2007, 2009). *Assessment Report.* Geneva: United Nations.

International Council for the Exploration of the Seas. (1990). *Marine Pollution Yearbook.* Oxford UK: Pergamon.

Jacobson, M. and M. Delucchi. (2009). "A Plan to Power 100 Percent of the Planet With Renewables." *Scientific American* (November).

Kahneman, D. (2011). *Thinking, Fast and Slow.* New York: Farrar, Straus and Giroux.

Kasperson, R. and J. Kasperson. (2005). *Social Contours of Risk: Volume II: Risk Analysis, Corporations and the Globalization of Risk.* London: Routledge.

Käuferle, J. (1972). "Using DC Links to Enhance AC System Performance." *IEEE Spectrum* 9 (June): 31–37.

Kennedy II, J. (2012). "The High Cost of Gambling on Oil." *The New York Times* (April 10).

Kerr, R. (2009). "Splitting the Difference between Oil Pessimists and Optimists." *Science* 326 (20 November): 1048.

Klare, M. (2005). *Blood and Oil.* New York: Holt.

Krauss, C. and E. Lipton. (2012). "U.S. Inches toward Goal of Energy Independence." *The New York Times* (March 22).

Krimsky, S. and A. Plough. (1988). *Environmental Risk.* Dover MA: Auburn House.

Kurokawa, K. (2012). *Fukushima Nuclear Accident Independent Investigation Commission.* Tokyo: National Diet of Japan.

Lambright, W. (2005) *Nasa and the Environment: The Case of Ozone Depletion.* Napa History Division: Washington, DC.

Lanouette, W. (1990). *Trutium and the Times.* Cambridge MA: Barone Center of Harvard University.

Lasker, G. (Ed.) (1981). *Applied Systems and Cybernetics.* New York: Pergamon.

Lefebvre, B. (2011). "Gas Stays High as Oil Drops." *Wall Street Journal* (October 1): A3.

Lee, J. (1993) "United States and British news coverage of oil spills, 1966–1990." Dissertation, Syracuse University, Syracuse, NY.

Leiss, W. (2001). *In the Chamber of Risks.* Montreal: McGill-Queen's University Press.

Lenin, V. (1920). *Collected Works,* Vol. 31, p. 516. Available online at http://www.marxists.org/archive/lenin/works/cw/index.htm

Lenski, G. (1970). *Human Societies.* New York: McGraw-Hill.

Lewis, S. (1922). *Babbitt.* New York: Harcourt, Brace & Co.

Licklider, R. (1988). "The Power of Oil: The Arab Oil weapon and the Netherlands, the United Kingdom, Canada, Japan, and the United States." *International Studies Quarterly* 32: 205–226.

Lichter, S. and S. Rothman. (1999). *Environmental Cancer: A Political Disease?* New haven CT: Yale University Press.

Lightfoot, H. (2006). "Understand the Three Different Scales for Measuring Primary Energy and Avoid Errors." *Energy* 32: 1478–1483.

Lippman, T. (2012). *Saudi Arabia on the Edge*. Washington DC: Potomac Books.
Lipton, E. (2012). "Even in Coal Country, the Fight for an Industry." *The New York Times* (May 29).
Loehr, G. (2007). "Enhancing the Grid: Smaller can be Better." *Energybiz Magazine* (January/February): 36–37.
Lovins, A. (1976). "Energy Strategy: The Road Not Taken?" *Foreign Affairs* 55 (October): 186–218.
Lovins, A. (2011). *Reinventing Fire*. White River Junction VT: Chelsea Green Publishing.
Mann, M. (2012). *The Hockey Stick and the Climate Wars*. New York: Columbia University Press.
Martin, P. (2005). *Twilight of the Mammoths*. Los Angeles CA: University of California Press.
Mazur, A. (1981). *The Dynamics of Technical Controversy*. Washington, DC: Communications Press.
Mazur, A. (1984). "The Journalists and Technology: Reporting about Love Canal and Three Mile Island." *Minerva* 22 (spring): 45–66.
Mazur, A. (1985). "Bias in Risk–Benefit Analysis." *Technology in Society* 7: 25–30.
Mazur, A. (1990). "Nuclear Power, Chemical Hazards, and the Quantity Of Reporting." *Minerva* 28 (autumn): 294–323.
Mazur, A. (1994). "How Does Population Growth Contribute to Rising Energy Consumption in America?" *Population and Environment* 15: 371–378.
Mazur, A. (1998). *A Hazardous Inquiry*. Cambridge MA: Harvard University Press.
Mazur, A. (2004). *True Warnings and False Alarms*. Washington DC: Resources for the Future.
Mazur, A. (2007). *Global Social Problems*. New York: Rowman & Littlefield.
Mazur, A. (2008). *Implausible Beliefs in the Bible, Astrology and UFOs*. New Brunswick, NJ: Transaction Publishers.
Mazur, A. (2009). "American Generation of Worldwide Environmental Warnings: Avian Influenza and Global Warming." *Human Ecology Review* 16: 17–26.
Mazur, A. (2011). "Does Increasing Energy or Electricity Consumption Improve Quality of Life in Industrial Nations?" *Energy Policy* 39: 2568–2572.
Mazur, A. (2012). "Was Rising Energy and Electricity Usage in Industrial Nations (since 1960) Due More to Population Growth or to Other Causes?" *Human Ecology Review* 19: 50–57.
Mazur, A. and J. Lee. (1993). "Sounding the Global Alarm: Environmental Issues in the US National News." *Social Studies of Science* 23: 681–720.
Mazur, A. and T. Metcalfe. (2012). "America's Three Electric Grids: Are Efficiency and Reliability Functions of Grid Size?" *Electric Power Systems Research* 82: 2568–2572.
Mazur, A. and E. Rosa. (1974). "Energy and Life-Style." *Science* 186 (14 November): 607–610.
McCombs, M. and D. Shaw. (1972). "The Agenda-Setting Function of Mass Media." *Public Opinion Quarterly* 36: 176–187.
McCright, A. and R. Dunlap. (2011). "The Politicization of Climate Change and Polarization in the American Public's Views of Global Warming." *Sociological Quarterly* 52: 155–194.
McGillivray, M. (2006). *Human Well-Being: Concept and Measurement*. London: Palgrave Macmillan.
McKinnon, A. (2007). "For an 'Energetic Sociology, or, Why Coal, Gas, and Electricity Should Matter for Sociological Theory." *Critical Sociology* 33: 345–356.
McLean, B. and P. Elkind. (2003). *The Smartest Guys in the Room*. New York: Penguin.
McNichol, T. (2006). *AC/DC: The Savage Tale of the First Standards War*. San Francisco CA: John Wiley.

McPhee, J. (1971). *Encounters with the Archdruid*. New York: Farrar, Straus and Giroux.

Michener, H. (1924). "Transmission at 220." *Transactions A.I.E.E.* (October): 1222–1225.

Mol, A., D. Sonnenfeld and G. Spaargaren. (2009). *The Ecological Modernisation Reader*. London: Routledge.

Morton, Jr., D. (2002). "Reviewing the History of Electric Power and Electrification." *Endeavour* 26 (2): 60–63.

Moyer, M. (2010). "The Dirty Truth about Plug-in Hybrids." *Scientific American* 303 (July): 54–55.

Muller, R. (2012). "The Conversion of a Climate-Change Skeptic." *The New York Times* (July 28).

Munson, R. (2005). *From Edison to Enron*. Westport CT: Praeger.

NAS. (2002). *Effectiveness and Impact of Corporate Average Fuel Economy (CAFE) Standards*. Washington DC: National Academy of Sciences.

Nelkin, D. (1979). *Controversy: Politics of Technical Decisions*. UK and US: Sage Publications.

New York City. (2012). *New York City Local Law 84 Benchmarking Report*. New York: Mayor's Office of Long-Term Planning & Sustainability.

Nordhaus, T. and M. Shellenberger. (2007). *Break Through: From the Death of Environmentalism to the Politics of Possibility*. Boston MA: Houghton and Mifflin.

Nye, D. (1992). *Electrifying America*. Cambridge MA: MIT Press.

Nye, J. (2001). *Consuming Power*. Cambridge MA: MIT Press.

Oreskes, N. and E. Conway. (2011). *Merchants of Doubt: How a Handful of Scientists Obscured the Truch on Issues from Tobacco Smoke to Global Warming*. New York: Bloomsbury Press.

Owen, D. (2009). *Green Metropolis*. New York: Penguin.

Owen, D. (2011). *The Conundrum*. New York: Penguin.

Pachauri, R. and A. Reisinger (Eds.) (2009). *Climate Change 2007 Synthesis Report. Geneva, Switzerland: IPCC*. Available online at http://www.ipcc.ch/publications_and_data/ar4/syr/en/contents.html

Park, H. (2010). *The Social Structure of Large-scale Blackouts*. PhD dissertation. Rutgers University.

PennWell. (1994). *Electric Power Generation and Transmission Systems Map of the US*. Tulsa: PennWell Publishing Company.

Perrow, C. (1999). *Normal Accidents*. Princeton NJ: Princeton University Press.

Petroski, H. (1992). *To Engineer Is Human: The Role of Failure in Successful Design*. New York: Vintage.

Palfreman, J. (2006). "A Tale of Two Fears." *Review of Policy Research* 23: 23–43.

Pimentel, D. and T. Patzek. (2005). "Ethanol Production Using Corn, Switchgrass, and Wood; Biodiesel Production Using Soybean and Sunflower." *Natural Resources Research* 14: 65–75.

Platt, H. (1991). *The Electric City*. Chicago: University of Chicago Press.

Platts. (2009/2010). *Transmission System of North America*. New York: McGraw-Hill.

Pomeranz, K. (2001). *The Great Divergence: China, Europe, and the Making of the Modern World Economy*. Princeton NJ: Princeton University Press.

President's Commission. (1979). *Report of the Public's Right to Information Task Force*. Washington DC: U.S. Government Printing Office.

Price, M. (2002). *Media and Sovereignty*. Cambridge, MA: MIT Press.

Proctor, R. (2012). *Golden Holocaust*. Berkeley CA: University of California Press.

Rasmussen, N., et al. (1975). *Reactor Safety Study*. Washington DC: Nuclear Regulatory Commission.

Revesz, R. and M. Livermore. (2008). *Retaking Rationality*. New York: Oxford University Press.

Rosa, E., A. Diekmann, T. Dietz and C. Jaeger. (2009). *Human Footprints on the Global Environment*. Cambridge MA: MIT Press.

Rosa, E., K. Keating and C. Staples. (1980). "Energy, Economic Growth and Quality of Life." In G. Lasker (Ed.) *Applied Systems and Cybernetics* (pp. 258–264). New York: Pergamon.

Rosa, E., G. Machlis and K. Keating. (1988). "Energy and Society." *Annual Review of Sociology* 14: 149–172.

Ross, M. (2012). *The Oil Curse: How Petroleum Wealth Shapes Development of Nations*. Princeton NJ: Princeton University Press.

Rothman, S., A. Kelly-Woessner and M. Woessner. (2011). *The Still Divided Academy*. Lanham MD: Rowman & Littlefield.

RWE. (2011). *Lignite – A Domestic Energy Source*. Cologne: RWE Power AG.

Sandman, P. and Paden, M. (1979). "At Three Mile Island." *Colombia Journalism Review* (July/August): 43–58.

Schipper, L. and A. Lichtenberg. (1976). "Efficient Energy Use and Well-being: The Swedish Example." *Science* 194 (3 December): 1001–1013.

Schipper, L. and S. Meyers. (1992). *Energy Efficiency and Human Activity*. New York: Cambridge University Press.

Schipper, M. (2006). *Energy-Related Carbon Dioxide Emissions in U.S. Manufacturing*. Washington DC: EIA.

Schneider, S. (1989). *Global Warming*. San Francisco CA: Sierra Club Books.

Seymour, L. and D. Kauneckis. (2012). *What is Preventing Geothermal Development in the Western United States?* Reno NV: University of Nevada.

Shabecoff, P. (2003). *A Fierce Green Fire*. Washington DC: Island Press.

Shoemaker, P. and A. Cohen. (2005). *News Around the World*. London: Routledge.

Singer, B. (1988). "Power Politics." *IEEE Technology and Society Magazine* (December): 20–27.

Singer, E. and P. Endreny. (1994). *Reporting on Risk*. New York: Russell Sage Foundation.

Slovic, P. (2011). *The Feeling of Risk: New Perspectives on Risk Perception*. London: Routledge.

Smil, V. (1994). *Energy in World History*. Boulder CO: Westview Press.Smil, V. (2003). *Energy at the Crossroads*. Cambridge, MA: MIT Press.

Smil, V. (2003). *Energy at the Crossroads*. MIT Press, Cambridge, MA.

Smil, V. (2005). *Creating the Twentieth Century: Technical Innovations of 1867–1914*. New York: Oxford University Press.

Smil, V. (2009). "U.S. Energy Policy: The Need for Radical Departures." *Issues in Science and Technology* (summer).

Sorrell, S. (2009). "Jevons' Paradox Revisited: The Evidence for Backfire from Improved Energy Efficiency." *Energy Policy* 37: 1456–1469.

Sovacool, B. and M. Brown. (2010). "Twelve Metropolitan Carbon Footprints: A Preliminary Comparative Global Assessment." *Energy Policy* 38: 4856–4869.

Springer, V. (1976). *Power and the Pacific Northwest*. U.S. Department of the Interior.

Starr, C. (1969). "Social Benefit Versus Technological Risk." *Science* 165; 1232–1238.

Starr, C. (1972). "Energy, Power and Society." *Scientific American* (September).

Stevens, W. (1989). "Split Forecast: Dissent on Global Warming." *The New York Times* (December 13).

Stokes, D. and S. Raphael. (2010). *Global Energy Security and American Hegemony*. Baltimore MA: Johns Hopkins University Press.

Swift, E. (2011). *The Big Roads*. New York: Houghton Mifflin Harcourt.

Tagliabue, J. (2010). "Norway Power Plan Upsets Nature Lovers." *The New York Times* (November 10).
Tierney, J. (1990). "Betting the Planet." *The New York Times Magazine* (December 2).
Time. (1971). "Environment: A Lemon Named Big Allis." *Time Magazine* (July 19).
Timmons, H. and M. Vyawahare. (2012). "India's Air the World's Unhealthiest, Study Says." *The New York Times* (February 1).
Troy, A. (2012). *The Very Hungry City*. New Haven CT: Yale University Press.
Van Ginneken, J. (1998). *Understanding Global News*. London: Sage.
Vaughn, D. (1997). *The Challenger Launch Decision*. Chicago: University of Chicago Press.
Von Daniken, E. (1968). *Chariots of the Gods*. US :Berkley Books.
Warburg, P. (2012). *Harvest the Wind*. Boston MA: Beacon Press.
Wald, M. (2002). "U.S., Alarmed by Corrosion, Orders Checking of Reactors." *New York Times* (March 26).
Ward, R. (2006). "Demographic Trends and Energy Consumption in European Union nations, 1960–2025." *Social Science Research* 36: 855–872.
Warkentin, D. (1998). *Electric Power Industry in Nontechnical Language*. Tulsa OK: PennWell.
Weare, C. (2003). *The California Electricity Crisis*. San Francisco CA: Public Policy Institute of California.
Weart, S. (2012). *The Rise of Nuclear Fear*. Cambridge MA: Harvard University Press.
Wellstone, P. and B. Casper. (2003). *Powerline*. Minneapolis MN: University of Minnesota Press.
White, L. (1959). *The Evolution of Culture*. New York: McGraw-Hill.
Wright, L. (2007). *The Looming Tower*. New York: Vintage.
Yergin, D. (1991). *The Prize*. New York: Simon & Schuster.
Yergin, D. (2011). *The Quest*. New York: Penguin.
Yergin, D. and J. Stanislaw. (2002). *The Commanding Heights*. New York: Free Press.
Yetiv, S. (2012). *The Petroleum Triangle: Oil, Globalization, and Terror*. Ithaca NY: Cornell University Press.
York, R., E. Rosa and T. Dietz. (2003). "Footprints on the Earth: The Environmental Consequences of Modernity." *American Sociological Review* 68: 279–300.
Zelizer, B. (1992). *Covering the Body*. Chicago: University of Chicago Press.
Zweibel, B., J. Mason and V. Fthenakis. (2008). "A Solar Grand Plan." *Scientific American* (January).

INDEX

Note: Page numbers in **bold** type refer to **figures**.